Predictive Methods in Percutaneous Absorption

Gary P. Moss · Darren R. Gullick
Simon C. Wilkinson

Predictive Methods in Percutaneous Absorption

Springer

Gary P. Moss
The School of Pharmacy
Keele University
Keele, Staffordshire
UK

Darren R. Gullick
College of Pharmacy
University of Georgia
Athens, GA
USA

Simon C. Wilkinson
Medical Toxicology Centre, Institute of Cellular Medicine
Newcastle University
Newcastle upon Tyne
UK

ISBN 978-3-662-47370-2 ISBN 978-3-662-47371-9 (eBook)
DOI 10.1007/978-3-662-47371-9

Library of Congress Control Number: 2015941113

Springer Heidelberg New York Dordrecht London

Printed on acid-free paper

Springer-Verlag GmbH Berlin Heidelberg is part of Springer Science+Business Media
(www.springer.com)

Dedicated to the pioneering research in this field by:
Gordon Flynn
Richard Guy
Russell Potts

Foreword

As a major function of the skin is to be a protective barrier and stop the ingress of exogenous chemicals, it may seem strange to dedicate decades of research effort into understanding of how, and to what extent, molecules pass through the skin! However, this topic, and the areas covered by this volume, is vitally important in understanding the potential beneficial or harmful effects of dermal penetration. Significant progress has been made through this research in understanding the structure, function and physiology of skin as well as how these factors influence the passage of molecules into and across it. This research area now goes beyond the physical measurement of the passage of molecules through the skin to include modelling and computational simulation technologies to assist our understanding of dermal penetration, as well as the chemistry behind formulation science. This knowledge has various applications in the pharmaceutical, personal product, biocide, fine chemical and many other manufacturing industries. This volume goes a long way to capture and define the state of the art in the experimental determination and computational calculation of percutaneous absorption.

However, despite some excellent research, we are still lacking the tools to consistently and reproducibly assess experimentally, let alone predict, the amount of a chemical that will reach the systemic circulation following topical application. There are many reasons for this, and this volume gives a thorough account of the problems, why they are important, and possible solutions. A substantial part of the problem is the quality, or otherwise, of the historical data with which we are developing models. Many of the data were not intended for the purpose for which they are now being used, rather being investigative studies of drug permeation or for risk assessment. As such, they more than adequately served their original purpose, but should be used with caution for modelling. A second problem is that of the formulated product which is applied to the skin. Modelling works best when the data used relate to the pure substance applied neat to the skin (or at most, is applied as a saturated aqueous solution). We are only beginning to understand and model the effect that other chemicals—particularly formulation components—have on penetration, something that is highlighted in this book. Therefore, in order to take the science forward, the information provided and conclusions reached in this volume are vital to integrate into novel research programmes to answer questions such as "How can I reach a therapeutic dose of a drug when applied dermally?" or "What will be the risk of using this shampoo on a daily basis?"

This book has been written by experts in the field and will provide a valuable resource and starting point for all who wish to venture into this area or continue their study. This work especially fulfils the ambitions of Dr. Gary Moss who has been researching in this field for two decades, starting with his Ph.D. from Queen's University Belfast. Following from his background in topical formulation development, he then combined this knowledge of experimental outcomes with a newfound interest in computational modelling methods—this was, after all, not long after the seminal publications of the Flynn compilation of skin permeability coefficients and the first Potts and Guy model. Gary has led and continues to lead the way in driving the process of data compilation and applying these approaches to other membranes (e.g. polydimethylsiloxane, Silastic®). Extending the concepts, through a variety of experiences garnered from positions in both industry and academia, he has worked to apply twenty-first-century modelling concepts to these problems, with careful reference to data quality, consideration of formulation and the adoption of novel machine learning methods. I have known Gary for over 20 years, and I have admired his scientific contributions to the field of transdermal research. There are few academic counterparts that can boast his knowledge of skin structure and function, transdermal drug delivery and toxicology applied to the permeation of exogenous chemicals into and across skin. This experience will become apparent to the reader of this textbook.

Simon Wilkinson is a toxicologist focusing on skin absorption and dermal metabolism, and has a strong interest in methodological approaches in percutaneous absorption. This expertise underpins one of the key themes in this book, which is the role of experimental data—usually derived for other purposes—applied to the modelling of percutaneous absorption and how this impacts on the model quality but also its relevance beyond theoretical or academic spheres.

An interesting finding, in research conducted by Dr. Moss and Dr. Darren Gullick, was the development of our understanding of skin permeability as a nonlinear phenomenon, which led to the development of further investigations using the Gaussian process machine learning techniques and which has informed significantly this current volume. Perhaps one of the author's key findings to date was the development of evidence for the nonlinear nature of the relationship between the physicochemical properties of a molecule and its percutaneous absorption, which paved the way for further investigations using machine learning methods, which could be considered to be the future of predictive percutaneous absorption research.

The current edition is a timely addition to the literature, presenting and updating us with the state of the art of predictive percutaneous absorption.

Prof. David Jones
Pro Vice Chancellor and Professor of Biopharmaceutics
The School of Pharmacy, Queen's University Belfast
Belfast, Northern Ireland, UK

Acknowledgments

The authors would like to thank Prof. Mark Cronin, Liverpool John Moores University, for his advice and comments on this book at various stages of its preparation.

Contents

Chapter 1
Skin Structure and Physiology

Introduction

The skin is the most complex and diverse organ in the human body. It has a multitude of important roles, which include temperature regulation, mechanical resistance and various protective functions—for example, both from the ingress of exogenous materials, chemicals and biological species, and the egress of water from the body.

The skin is the largest organ of the body. On average, it accounts for approximately 10 % of body mass, receives approximately one-third of the blood circulating throughout the body and has a surface area of approximately 2–3 m^2 (Woolfson and McCafferty 1993a, b). It provides a strong yet flexible self-repairing barrier to the external environment and protects internal body organs and fluids from external influences, harmful molecules and micro-organisms. The skin also forms an extensive sensory surface, transmitting sensations such as heat, cold, touch, pressure and pain to the central nervous system. The skin is a multilayered organ consisting of three main histological layers: the epidermis, the dermis and the subcutis. Mammalian skin is a stratified epithelium, and each layer will be considered individually, below, progressing from the deeper (innermost) tissues to the outermost tissues (those outermost tissues which are, ultimately, in contact with the external environment).

The Hypodermis (Subcutaneous Fatty Tissue Layer)

At the base of the skin, below the epidermis and dermis, lies the subcutaneous fatty tissue layer, often called the subcutis, or hypodermis. It provides support and cushioning for the overlying epidermal and dermal layers, a means of attachment to

G.P. Moss et al., *Predictive Methods in Percutaneous Absorption*,
DOI 10.1007/978-3-662-47371-9_1

deeper tissues below the skin. It acts as a depository for fat and an absorber of external forces, such as heat and physical shock, and contains the blood vessels that supply the skin. It is variable in thickness depending on the body site; it ranges from a few centimetres thick in some regions (i.e. the abdominal wall) to areas where there is little or no fat and where the hypodermal layer may be difficult to observe (i.e. the eyelid or the scrotum). As the dermis and hypodermis are both irregular connective tissues, it is often difficult to distinguish between them. However, the hypodermis is generally looser and contains a higher proportion of adipose cells compared with the dermis.

The Dermis

The dermis (or corium) lies immediately above the hypodermis. In terms of volume, it is the largest part of the skin, being commonly ten to twenty times thicker than the epidermis. It is usually 0.1–0.5 cm in thickness, depending on its location. It is a robust and durable tissue that provides flexibility and tensile strength to the skin. Its myriad functions include protecting the body from injury and infection and provision of nutrition for the epidermis. It also contains the main source of water within the skin. The dermis is comprised mostly of collagen, arranged in mechanically strong fibrous chains, which sit within a mucopolysaccharide gel-like structure (Wilkes et al. 1973). This matrix hosts a range of important structures, including nerve tissues, vascular and lymphatic systems, and the bases of various skin appendages. The lower part so the dermis consists of coarse fibrous tissues which provide the main supporting structural layer of the skin. It is also the locus of the blood vessels, which may extend to within 0.2 m of the skin surface (Woolfson and McCafferty 1993a, b). Towards the top of the dermis, the connective structures are more loosely formed and contain a finely structured papillary layer which encroaches into the epidermis. The transition between the predominately fibrous dermal tissues and the predominately cellular epidermal layers occurs at the dermo-epidermal junction.

The vasculature system of the skin is responsible for regulation of skin temperature, the supply of nutrients and oxygen to the skin, and the removal of toxins and waste products in assisting wound repair and healing. In the context of percutaneous absorption, it plays an important role in the removal of locally absorbed chemicals by carrying them into the systemic circulation. As the skin's blood supply can become relatively close to the skin's surface, penetrants are therefore removed from the skin at around the dermo-epidermal junction. This implies that the lower dermal and hypodermal layers of the skin play little role in the process of percutaneous absorption. It also implies that the blood supply to the skin provides the opportunity for penetrants to be removed from the local tissues and hence is important in the maintenance of a concentration gradient across the skin barrier.

Cross and Roberts (1993) also commented that the lymphatic system, which is located a comparable distance from the exterior of the body, may also play a significant role in the clearance of exogenous penetrants.

Skin Appendages

Human skin has associated with it several types of appendages, including hair follicles and their associated sebaceous glands (Fig. 1.1), and eccrine and apocrine sweat glands.

On average, human skin contains 40–70 hair follicles and 200–250 sweat ducts per square centimetre of skin. The skin appendages occupy approximately 0.1 % of the total skin surface, although this varies from region to region with, for example, the axillary, anogenital area and forehead having a larger than average concentration of hair follicles (Bronaugh and Maibach 1999). Hairs are formed from compacted plates of keratinocytes and reside in the hair follicles, which are invaginations in the epidermis. Sebaceous glands are associated with the hair follicles—usually formed as outgrowths of the follicle. They secrete an oily material, sebum, onto the skin surface. Sebum is a lipid-rich mixture which acts as a plasticiser for the *stratum corneum* and helps to maintain an acidic mantle of approximately pH 5 (Bronaugh and Maibach 1999). Eccrine glands are principally concerned with temperature control and are responsible for the secretion of sweat when stimulated by an increase in the external temperature or emotional factors. These glands commonly occupy approximately 10^{-4} % of the total skin area, and their structures ensure that they extend well into the dermis. Eccrine glands are found throughout the body, while apocrine glands are located in specific regions, including the axillae and anogenital regions.

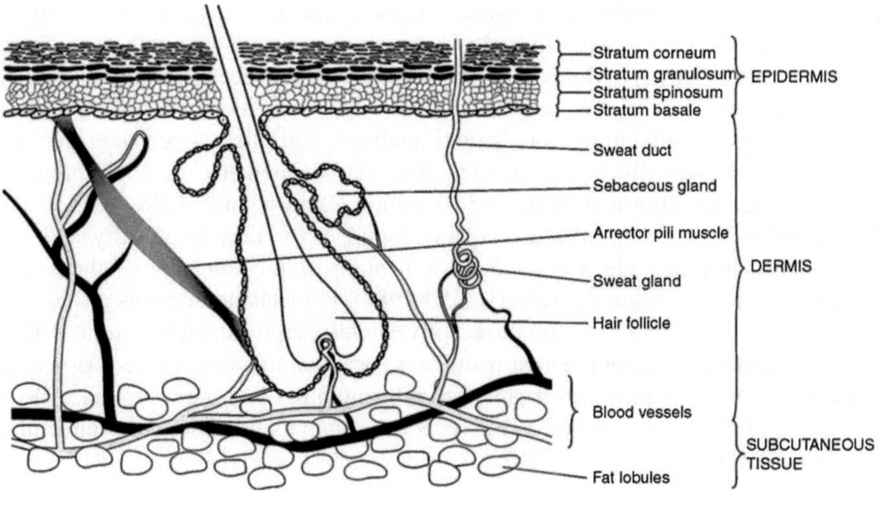

Fig. 1.1 Schematic diagram of the skin (© Williams (2003), used with permission)

The Subcutaneous Sensory Mechanism

The large size of the skin means that it acts as a major sensory organ for the body, particularly as it interfaces with the external environment. It provides information about the environment directly and indirectly, such as the effect of radiation on skin temperature. Fibres within the dermis form a plexus which lies parallel to the skin surface. The nerve plexus is comprised of unmyelinated and myelinated fibres. From the nerve plexus, individual fibres extend to supply particular locations in terminal branches which interconnect with and superimpose themselves upon each other in such a manner that every area in the skin is supplied by several different fibres, each of which ends in at least one particular receptor (Weddell 1941). Most of these receptors can be excited by different stimuli, but the different thresholds of stimuli required to provoke a particular receptor yield its specificity (Barlow and Mallon 1982).

The three main categories of cutaneous receptor, which are distinguished by their different sensitivities to stimuli, are the mechanoreceptors, thermoreceptors and nociceptors.

Mechanoreceptors are highly sensitive to pressure on the skin, or to movement of the hairs. Mechanoreceptors are usually described as rapidly adapting (RA) or slowly adapting (SA) types. RA mechanoreceptors include Pacinian corpuscles, which are found in both hairy and glabrous skin, and Meissner's corpuscles, which are located in the glabrous skin of primates. Pacinian corpuscles are small oval structures found in the deeper layers of the skin. They are 0.5–2 mm long and are composed of an "onion-like" lamellar structure which is formed from non-nervous tissue. Pacinian corpuscles are able to detect mechanical vibrations at high frequencies, which may be relayed at greater than one hundred hertz per second (Brodal 1981; Sinclair 1981). The Meissner corpuscle is an encapsulated myelinated receptor found in the dermis of human glabrous skin. It is surrounded by connective tissue which is attached to the basal projections of the epidermal cells by elastin fibrils. The Meissner corpuscle allows discrimination between highly localised sensations of touch, especially in the palmar regions where they are found in their highest density (Montagna 1964). Hair follicle receptors are myelinated fibres which are primarily associated with the tactile sensations (Elliott 1969). SA mechanoreceptors, including the Ruffini endings and the C-mechanoreceptors, respond during skin displacement, including the maintenance of a discharge of impulses when the skin is held in a new position (Barlow and Mallow 1982). The Ruffini endings are encapsulated receptors found in the dermis of hairy and glabrous skin. They provide a continuous indication of the intensity of the steady pressure or tension within the skin (Brodal 1981). C-mechanoreceptors are usually found in hairy skin and have small receptive fields (approximately 6 mm^2). They emit a SA discharge when the skin is indented or when hairs are moved. However, repetitive stimulation produces a rapid fall in excitability and the receptors will fail to respond after 20–30 seconds as the receptor terminals become unexcitable after this time (Barlow and Mallon 1982).

Thermoreceptors are characterised by a continuous discharge of impulses at a given constant skin temperature which increases or decreases when temperature is raised or lowered. Thermoreceptors have small receptive fields (approximately 1 mm^2) and are classed as either "cold" or "warm" receptors, with the former located more towards the outer surface of the skin than the latter, at average depths of 0.15 and 0.6 mm below the skin surface, respectively (Barlow and Mallon 1982). While thermo- and mechanoreceptors contribute to the sensory quality of perceived pain, the nociceptors detect and signal high intensities of stimulation. Nociceptors generally reside at the dermo-epidermal junction and are either mechanical nociceptors (which respond to, for example, pinpricks or needles, or squeezing and crushing of the skin) or thermal, or mechanothermal, nociceptors (which respond to severe mechanical stimuli and to a wide range of skin temperatures) (Brodal 1981; Montagna 1964).

The Epidermis

The epidermis is the outermost layer of the skin. It is also the thinnest layer of the skin. Its thickness varies significantly around the body with, for example, the thickest skin being found on the weight-bearing planter surfaces (feet and hands, approximately 0.8 mm thick) and the thinnest skin being normally found on the eyelids and scrotum (0.06 mm) (Williams 2003). Despite the extensive vasculature present in deeper tissues such as the dermis, the epidermis has no blood supply and passage of materials into or out of it is usually by a process of diffusion across the dermo-epidermal junction and into the dermis. It is essentially a stratified epithelium, consisting of four, or often five, distinct layers (Fig. 1.2).

The Stratum Germinativum

The *stratum germinativum*, or basal layer, is the deepest layer of the epidermis. This metabolically active layer contains cells similar to those found in other tissues in the body and contains organelles such as mitochondria and ribosomes. It can be as thin as a single cell in depth and contains cuboid or columnar-to-oval-shaped cells which sit on the basal lamina. These cells are continually undergoing mitosis, as they provide replacement cells for the higher (outer) epidermis. Basal keratinocytes are connected to the dermo-epidermal membrane by hemidesmosomes and connect the basal cells to the basement membrane. The basal layer is also the location of other cells, including melanocytes, Langerhans cells and Merkel cells. The basal cells become flatter and more granular as they move up through the epidermis.

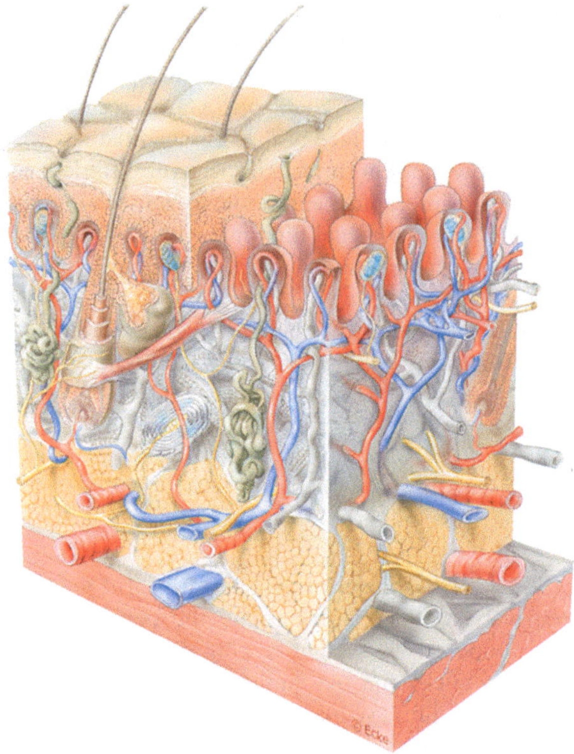

Fig. 1.2 Schematic representation of the epidermis (*source* BASF Personal Care and Nutrition GmbH; available at http://www.skin-care-forum.basf.com)

The Stratum Spinosum

The *stratum spinosum*, or prickle cell layer, sits immediately above the *stratum germinitivum*. It is often described with the basal layer (where the basal layer is very thin) as a single layer—the Malpighian layer. Normally, however, it is several layers thick (usually 2–6 layers) and consists of morphologically irregular cells which may range from columnar to polyhedral in structure; such a progression in morphology is common as this layer progresses upwards. Each cell in this layer possesses tonofilamental desmosomes, often called "prickles" or "spines", which give this layer its characteristic name and extend from the surface of the cell in all directions, helping to maintain a distance of approximately 20 nm between cells. The prickles of adjacent cells link via intercellular bridges and give three-dimensional structural rigidity and increase the resistance of the skin to abrasion throughout this layer. The prickle cell layer is metabolically active despite lacking in mitosis.

The Stratum Granulosum

The *stratum granulosum*, or granular layer, lies immediately above the *stratum spinosum* and is usually one to three cells deep. It consists of flattened, granular cells whose cytoplasm contains characteristic granules of keratohyalin, which is responsible for their characteristic "granular" appearance. In the *stratum granulosum*, degradation of cell components becomes significant; visually, this is seen in the flattening of cells compared to the layers immediately below the granular layer, and also in a substantial decrease in metabolic activity which eventually ceases towards the top of this layer due to the degeneration of cell nuclei, which leaves them unable to carry out important metabolic reactions.

The Stratum Lucidum

The *stratum lucidum* sits immediately above the *stratum granulosum.* It is easily observed on thick skin, but may be missing from thinner skin, which is why the epidermis is often described as having either four or five layers. The *stratum lucidum* is often considered to be functionally indistinct from the *stratum corneum* and that it may be an artefact of tissue preparation and cell differentiation, rather than a morphologically distinct layer. The cells of the *stratum lucidum* are elongated, translucent, and mostly lack either nuclei or cytoplasmic organelles. This layer is significantly more keratinised, and contains significantly flatter cells, than the underlying layers of the epidermis.

The Stratum Corneum

The outermost layer of the skin is the *stratum corneum*, or horny layer. It is the final result of cell differentiation and compaction prior to desquamation and removal from the body. It is a compacted, keratinised multilayer which is dehydrated in comparison with the adjacent layers of the skin. It is, on average, 15–20 cells thick —around 10 μm in thickness when dry, although it can swell to many times its thickness when wet. The formation of keratin and the resultant cell death are part of the process of keratinisation or cornification that produces what is, in effect, the *stratum corneum*, the outer envelope of the body. In areas of the skin where the *stratum lucidum* is clearly present, the *stratum corneum* is usually much thicker, and this also mirrors the thickness of the viable epidermis around the body. Thus, the epidermis in those regions, such as the palms and soles, can be up to 800 μm in thickness, compared to 75–150 μm in other areas. Cells of the *stratum corneum* are physiologically inactive, continually undergoing a process of shedding while themselves being constantly replenished from the upward migration of cells from the underlying epidermal layers (Woolfson and McCafferty 1993a, b).

The *stratum corneum* is the major rate-limiting membrane of the skin and is responsible for the regulation of water loss from the body as well as limiting the ingress of harmful materials from the external environment (Scheuplein and Blank 1971). It is currently believed to consist of two alternating amorphous lipophilic and hydrophilic layers and is comparatively more lipophilic than the other epidermal layers. While generally having lower water content than other layers of the skin, the *stratum corneum* water content is highly variable and depends on both moisture content of the external environment of the body and the location on the body. The exceedingly dense *stratum corneum* may also swell to many times its own thickness in the presence of water. The water content of the *stratum corneum* generally decreases as the external interface is approached. The *stratum corneum* possesses approximately 40 % water by weight (in a relative humidity of 33–50 %). By weight, the *stratum corneum* is composed of approximately 40 % protein (mostly keratin) and 15–20 % lipid (triglycerides, cholesterol, fatty acids and phospholipids) although it should be noted that the exact composition will vary around the body. The *stratum corneum* lipids originate from a number of sources, including the discharged lamellae of membrane-coated granules, intercellular cement and the keratinocyte cell envelope (Anderson and Cassidy 1973).

The cells of the *stratum corneum* are flattened and elongated and are approximately 1 μm in thickness. They occupy an area of 700–1200 μm^2; thus, there are approximately 10^5 cells per cm^2. They form a closely packed array of interdigitated cells (which facilitates the formation of cohesive laminae) which are the cells stacked in vertical columns (MacKensie and Linder 1973). Each cell is contained within a mainly proteinaceous envelope rather than the conventional lipid bilayer cell membrane. This envelope provides the *stratum corneum* with the majority of its mechanical strength, in particular through the disulphide bonds of the intracellular keratin and by linking cells that are embedded in an intercellular lipid matrix (Matolsty 1976).

The upward movement of cellular material through the epidermis ends in the *stratum corneum*, which constantly sheds its outermost layers in a process called desquamation. This process involves the cleavage of intercellular bridges and may suggest a certain amount of metabolic activity and regulatory control in a layer often considered to be, to all intents and purposes, inert (Michaelis et al. 1975). Typically, the daily desquamatory loss from the *stratum corneum* is typically not more than 1 g, although as the rate of *stratum corneum* shedding is, in healthy skin, equal to the rate of epidermal cell regeneration, the *stratum corneum* remains approximately the same thickness.

The Stratum Corneum *Barrier*

The *stratum corneum* skin barrier has traditionally been described as a "bricks and mortar" structure (Michaelis et al. 1975; Elias et al. 1981). The "bricks" represent the tightly packed corneocytes, and they are embedded in a "mortar" of lipid bilayers. These flattened, highly proteinaceous cells are the final point of keratinocyte differentiation and are interconnected by structures called corneodesmosomes (Fig. 1.3). The "bricks" are enclosed within a continuous and highly ordered lamellar lipid bilayer. Structurally, ceramides are the most important components of this lamellar phase; they are polar lipids which contain hydroxylated alkyl side chains that, under normal conditions, are packed both hexagonally and orthorhombically. As discussed above, the keratinocyte "bricks" of the skin barrier may hydrate extensively, resulting in significant changes to the packing, structure and permeability of the *stratum corneum* (Michel et al. 1988; Norlen 2006, 2007; Rawlings 2003, 2010). The *stratum corneum* corneocytes change in their morphological and biochemical functions as they progress from the lower to higher levels of the *stratum corneum.* Such transitions are associated with increases in transglutaminase-mediated protein crosslinking and increased levels of intercorneocyte ceramides and fatty acids, resulting in a progression from fragile to rigid structures, described by Rawlings (2010) as the transition from "*stratum compactum*" to "*stratum disjunctum*". This transition occurs alongside an increase in the occurrence of the protein (pro)filaggrin, which is thought to play a key role in the aggregation of keratin filaments within corneocytes (Rawlings 2010).

Despite the fundamental correctness of the "bricks and mortar" model of this *stratum corneum,* advances have been made in recent years, which have elaborated our understanding of the *stratum corneum* structure and its barrier function. New species of ceramides and the synthetic pathways that generate them are still being identified (Rawlings 2010). Cryoelectron microscopy studies have proposed the existence of a

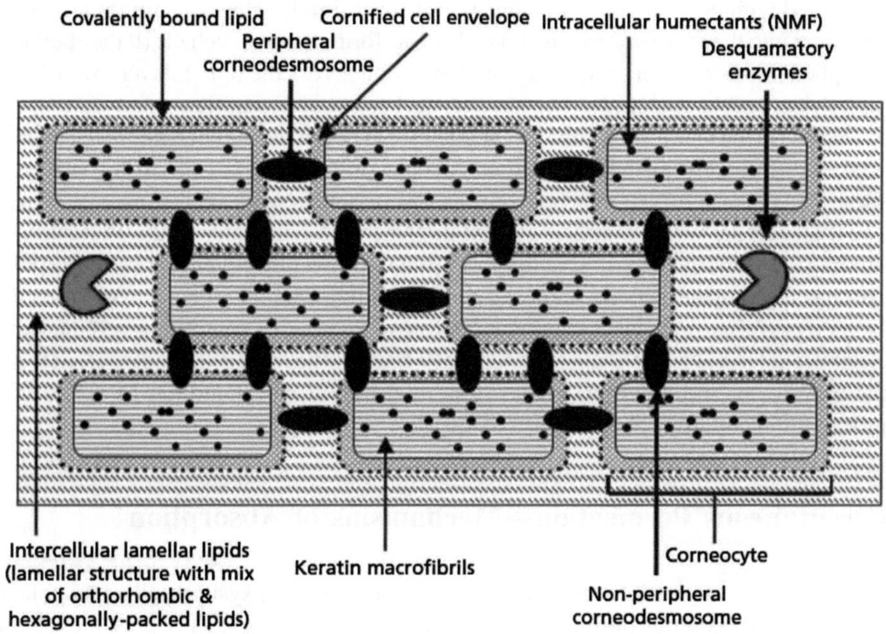

Fig. 1.3 Schematic structure of the *stratum corneum* (Rawlings 2010, used with permission)

single gel phase model for the *stratum corneum* lipids while failing to show the expected presence of the trilamellar-conformation long periodicity phase (Norlen 2007). Bouwstra et al. (1998) suggested that the *stratum corneum* lipid phase could be described by a "sandwich model" which explains differences observed in *stratum corneum* lipid packing, particularly with regard to differing periodicity phases. This model highlights the importance of a fluid phase within the *stratum corneum* which may be dictated by the presence of ω-esterified long-chain acylceramides.

Ultimately, the *stratum corneum* cannot be considered as a homogenous tissue as it exhibits clear changes as it progresses outwards from the body—the transition from "*compactum*" to "*disjunctum*". This transition may be exemplified by a transition in the packing of ceramide side chains from a more tightly packed orthorhombic phase to a less tightly packed hexagonal phase which becomes increasingly present closer to the skin surface. In addition, at the skin surface the lamellar phase is normally missing as it becomes increasingly amorphous in nature at this point (Pilgram et al. 1999, 2001; Rawlings 2010).

Routes of Permeation of Exogenous Chemicals Across the *Stratum Corneum*

It is widely understood that the main route for exogenous chemicals to pass into and across the *stratum corneum* is via the lipid pathway. Despite being a longer and more torturous route across this layer, it does not require the multiple partitioning steps associated with the transcellular pathway. Rather, it simply relies on partitioning of the penetrant into the *stratum corneum* lipids from its formulation or vehicle (if the chemical is applied in this manner) and subsequent diffusion across the lipid bilayer towards the underlying viable epidermis. The appendageal route of absorption—permeation, for example, via the hair follicles and sweat glands—is of limited significance in the overall permeation process as these structures occupy, on average, 0.1 % of the total skin surface and therefore provide a limited target for permeation. Further, structures such as sweat glands are often morphologically similar to the remainder of the skin surface, limiting the viability of these structures as a route of absorption even more as absorption also has to compete with an opposing outward flow of liquid when active. Thus, the *stratum corneum* lipids play the dominant role in limiting or controlling percutaneous absorption. However, it should be noted that while it is the most important route, it is not the only route, and that other routes of administration will contribute to the overall process of skin permeation (Moss et al. 2012).

Percutaneous Permeation—Mechanisms of Absorption

For a chemical to pass into and across the skin, and become systemically available, it must undertake a series of partitioning steps. The chemical is usually presented to the skin surface in a formulation or vehicle from which it must partition onto the skin

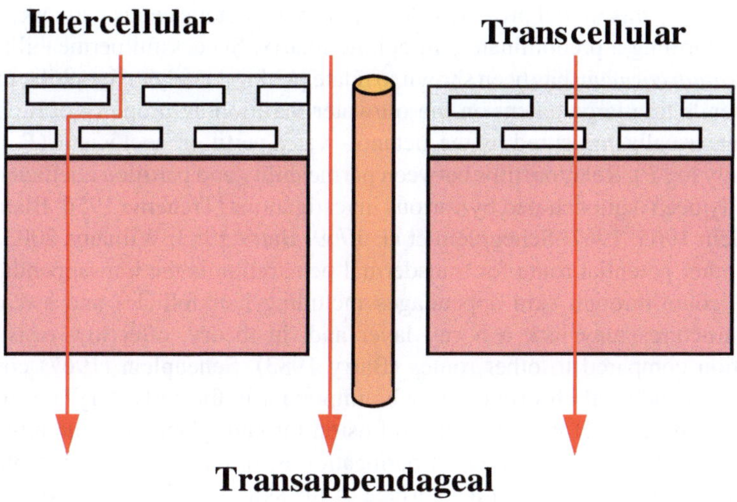

Fig. 1.4 Pathways of drug penetration through skin

surface, where those molecules in contact with the *stratum corneum* will begin to partition. Depending on their properties, the nature of both the vehicle or formulation and the penetrant can play a key role in determining the rate at which the chemical penetrates the skin. For example, if the penetrant has a high affinity for the formulation, then it may remain there, whereas if it has a low affinity for the formulation (or a higher affinity for the *stratum corneum*), then it may partition into the skin more readily. Thus, the penetrant adjacent to the skin surface will permeate into the *stratum corneum*, a process which is dependent on the random movement of the penetrant from the bulk of the vehicle to the surface of the skin, which again may be influenced by the nature of the vehicle or formulation. Once the penetrant has diffused into the *stratum corneum,* it will begin to diffuse through this layer, with the rate of diffusion again depending on the physicochemical properties of the penetrant. This may occur via any of the three main routes described above (intracellular, intercellular and transappendageal; shown in Fig. 1.4). Permeation may be either via a specific route or a combination of any of the available routes. The next significant challenge to permeation is at the junction of the *stratum corneum* and the viable epidermis. At this point, the underlying tissues may be broadly differentiated from those above as they are more hydrophilic than the outer layers of the epidermis, and the *stratum corneum* in particular. This results in a further partitioning step and diffusion into the viable epidermis, therefore partitioning between the viable epidermis and the dermis. Finally, partitioning from the dermis to the capillary system results in the penetrant being removed to the systemic circulation.

The transepidermal route, via the intact *stratum corneum*, is the main route through which penetrants may enter, as it provides the major area available to a potential penetrant. The *stratum corneum* has been morphologically and functionally represented by the "bricks and mortar" model (Elias 1988). The "bricks", or corneocytes, of this model provide a dense, fibrous, proteinaceous network, with the "mortar" forming a predominately lipophilic matrix. Successful permeability of the intact *stratum corneum* has been shown to relate predominately to lipophilic materials and depends to a large extent on the oil/water partitioning property of a particular penetrant, usually measured as the octanol–water partition coefficient (*P*, or more commonly log *P*). Relationships between permeability and partition coefficients have classically been demonstrated by various investigations (Treherne 1956; Blank 1964; Scheuplein 1965, 1967; Scheuplein et al. 1969; Barry 1983; Williams 2003).

The other potential route for transdermal penetration is the transappendageal, or "shunt", route through skin appendages including hair follicles and sweat ducts. These structures may lack a horny layer and, in theory, offer low resistance to permeation compared to other routes (Barry 1983). Scheuplein (1967) concluded that transappendageal absorption may be important in the early "lag" period of the penetration process. However, while diffusion through glands is generally considered to occur, the rate and extent of permeation is, in most cases, negligible due to the small area they occupy on the surface of the skin, and the current of secretions passing to the outer surface as mentioned earlier, which is often mediated by valve mechanisms at the openings of the glands (Barr 1962; Heuber et al. 1992, 1994). Ultimately, however, successful permeation of exogenous chemicals via the shunt route depends predominately upon the physicochemical properties of the penetrant as well as the nature of the *stratum corneum* and may be more successful for some penetrants than for others. In addition, other factors may influence the penetration process. For example, the potential for protein binding, which may occur in the *stratum corneum*, will contribute to the reservoir effect associated with that layer. Metabolic activity may see some, or potentially all, of the permeant degraded before it reaches the blood vessels. There is also potential for permeants to pass into deeper layers of the skin, including the subcutaneous fatty layer, or even into muscle tissues underlying the skin.

Theoretical Considerations

Diffusion is "a process of mass transfer of individual molecules of a substance, brought about by random molecular motion and associated with a concentration gradient" (Martin et al. 1983). Diffusion through a non-porous membrane, such as the *stratum corneum*, occurs when the diffusant dissolves in the bulk membrane or solvent-filled pores of the membrane. Such diffusion is influenced by the size and physicochemical properties of the penetrant and the nature of the membrane, and possibly also the formulation or vehicle particularly if it exerts a change on the nature of the membrane. While the three layers of the skin (the epidermis, the dermis and the subcutis) each have their own diffusion coefficient, diffusion through any layer other than the *stratum corneum* is generally considered to be negligible and, as such, they are normally treated together and represented by a single diffusion coefficient.

Thus, total diffusional resistance of the skin is generally attributed to the *stratum corneum* under passive diffusion, and therefore, Fick's first law of diffusion may be applied (Martin et al. 1983; Moss et al. 2002):

$$J = -D\frac{\partial C}{\partial x} \tag{1.1}$$

where

J is the rate of transfer per unit area of the surface (i.e. the flux);
C is the concentration of the diffusing substance;
x is the spatial coordinate measured normal to the section; and
D is the diffusion coefficient, or diffusivity.

The dermal permeability coefficient, *k*p, is defined by the equations:

$$J_{ss} = k_p C_v \tag{1.2}$$

or

$$k_p = J_{ss}/C_v \tag{1.3}$$

Combination of Eqs. (1.2) and (1.3) gives

$$k_p = K_m \cdot D/h \tag{1.4}$$

where

k_p is the permeability coefficient (cm/s or cm/h);
C_v represents the concentration of penetrant in the vehicle when sink conditions apply;
J_{ss} is the steady-state flux of the solute;
D is the average diffusion coefficient (cm^2/s or cm^2/h);
K_m represents the partition, or distribution, coefficient between the *stratum corneum* and the vehicle; and
h is the thickness of the skin.

Thickness of the membrane has generally been recognised as being inversely proportional to flux, although Elias et al. (1981) suggested that lipid content, rather than thickness, was of greater relevance. Further, the above steady-state model is more appropriate for in vitro systems, as it is unlikely to hold in more complex in vivo situations due to the low permeability of the *stratum corneum*. Nevertheless, in vitro diffusion is still a highly important area of research, particularly in the development of models for percutaneous absorption, providing excellent theoretical and preliminary investigative models of in vivo permeation for a range of endpoints, including pharmaceutical efficacy and safety/toxicity. Thus, from the viewpoint of the percutaneous absorption of exogenous chemicals into and across the skin, the *stratum corneum* is often considered to be essentially a simple lipid which interfaces with a predominately hydrophilic layer sitting immediately beneath it. The transport of lipophilic chemicals occurs predominately via the *stratum corneum*, and as these compounds must transfer directly from this comparatively lipid-rich environment into an aqueous medium, compounds that are highly lipophilic will remain largely in the *stratum corneum* or permeate at a very slow rate.

Physicochemical Properties of the Penetrant

Classically, the physicochemical properties of a penetrant are known to significantly influence its ability to penetrate into and across the skin; more broadly, this applies to the permeation by exogenous chemicals of a number of routes of administration or entry to the body. In the section below, the main chemical/molecular properties of penetrants will be considered in the context of skin absorption, normally as discrete parameters or descriptors of (often composite) molecular properties.

Partition Coefficient

The partition coefficient is the ability of a substance to partition between two immiscible phases, usually octanol–water or heptane/buffer. Somewhat simplistically, a higher partition coefficient represents a more lipophilic molecule and is usually associated experimentally with an increase in permeation via the lipid domains of the *stratum corneum*. For a chemical to cross the *stratum corneum,* it must first partition into this membrane, and this may be the rate-limiting step in the permeation process. Barry (1987) determined that the partition coefficient, usually described as log P or log K_{OW}, of a penetrant will influence the path it takes in traversing the skin. For example, Bronaugh and Congdon (1984) demonstrated that, for a series of hair dyes, increasing the lipophilicity of a molecule increased the rate of penetration, while Le and Lippold (1995) indicated that the maximum flux may be estimated from the penetrant's physicochemical properties, particularly the partition coefficient. Higo et al. (1995) demonstrated that skin penetration was dependant on the partition coefficient for a series of salicylic acid derivatives. Predominately hydrophilic permeants will have a comparatively higher tendency to permeate across the skin via hydrophilic pathways, such as hydrated keratin-filled keratinocytes. In this case, the effect of the partition coefficient for such penetrants is not as clear. For example, the lipid bilayer contains hydrophilic elements, such as the polar head groups of lipids, suggesting that hydrophilic permeants may traverse the skin barrier by a number of different routes. Williams suggested that permeants with intermediate properties—defined as having a log P of between 1 and 3—will traverse the skin barrier via both lipid and aqueous pathways but the intercellular route predominates (Williams 2003). Lipophilic molecules (those with a log *P* greater than 3) will predominately partition via the intercellular pathway. In practice, the ideal transdermal penetrant should possess both lipophilic and hydrophilic properties due to the predominately lipophilic nature of the *stratum corneum* and the increasingly hydrophilic nature of the underlying skin strata (Barry 1983; Sinko 2005).

Molecular Size and Shape

Consideration of the size and shape of a molecule is an important factor in determining its suitability as a percutaneous penetrant. While molecular volume is the most appropriate term to consider, molecular weight is more frequently used due to convenience and practicality (Williams 2003; Mitragotri et al. 2011). In general, an inverse relationship exists between the diffusivity of a molecule and its molecular weight, and as such small molecules may diffuse comparatively faster within a particular medium with a cut-off limit to absorption being generally associated with a molecular weight of 500 Da (Crank 1975; Idson 1975). Scheuplein and Blank (1969) compared the rates of penetration of a series of related compounds, all consisting of four carbon atoms and varying in the position of either one or two added oxygen atoms, which were present as various functional groups. They showed that permeability varies greatly when the functional groups are changed and that the least permeable molecules are those which are the most polar. Scheuplein and Blank also demonstrated that the skin permeability of steroids decreases when they are modified to incorporate more polar functionalities, such as hydroxyl groups.

Applied Concentration/Dose

Increasing the concentration of a chemical within a topically applied vehicle generally increases the amount of chemical absorbed across the skin (Maibach and Feldman 1969; Barry 1983; Williams 2003). Further, increasing the surface area available for permeation, within practicable limits, increases the potential for a topically applied molecule to be absorbed across the skin (Crank 1975; Wester and Noonan 1980; Sved et al. 1981). Frequency of application will also affect the delivered dose; although one large application usually results in the absorption of a higher dose, a single application may also have a greater toxicological potential compared to frequent, smaller doses (Wester et al. 1977, 1980; Wilson and Holland 1982). Occlusion and duration of contact can also increase the amount of applied chemical absorbed (Howes and Black 1976; Nakaue and Buhler 1976).

Aqueous Solubility and Melting Point

The percutaneous penetration of a molecule is greatly influenced by its aqueous solubility and partition coefficient. Lipophilic molecules generally penetrate the *stratum corneum* more rapidly than hydrophilic molecules. However, this needs to be balanced with preferential solubility in deeper layers of the viable epidermis and dermis. The partition of the penetrant between the *stratum corneum* and its vehicle or formulation is of great importance in percutaneous absorption. If the drug is more soluble in the *stratum corneum* than the vehicle, then the concentration of that chemical in the *stratum corneum* may be greater than in the vehicle at equilibrium. Where drugs are fully solubilised in the formulation, the rate of penetration is generally increased by complete diffusion in the vehicle and may be due to improved diffusion through the vehicle, which replenishes the vehicle/skin interface. Further, melting point is well correlated with aqueous solubility, to the extent that predictive models often employ melting point to determine solubility (Ostrenga et al. 1971a, b).

Ionisation

The predominately lipophilic nature of the *stratum corneum* and its largely lipophilic pathway suggests that the unionised form of a molecule is more likely to permeate the skin than the ionised form. The degree of penetrant ionisation is therefore essential in optimising the permeation of topically applied chemical, particularly drugs. According to the pH partition theory, if a molecule is unionised, then it may readily penetrate the *stratum corneum* via the intercellular pathway, as lipophilic regions of the skin act as barriers to ionised species and that ionised species may permeate the skin via the transappendageal route (Shore et al. 1957; Swarbrick et al. 1984). Parry et al. (1990) demonstrated—both experimentally and by the application of a mathematical model—that only unionised species enter and traverse the skin, while Roy and Flynn (1990) demonstrated that the unionised, free base forms of fentanyl and sufentanil are 218 and 100 times, respectively, more permeable than the ionised forms. They concluded that the contribution to the process of passive diffusion by ionised species is negligible.

Nevertheless, such comments should be taken in the wider context of a penetrant's physicochemical properties relative to the complex diffusive pathways available within the skin. Thus, a number of studies have shown that both ionised and unionised molecules can penetrate a lipophilic membrane, although the rates of transport and routes taken are significantly different for both species (Barker and Hadgraft 1981; Swarbrick et al. 1984; Siddiqui et al. 1985). For example, ionised compounds have been shown to penetrate the skin by mechanisms of either ion-pairing (Barker and Hadgraft 1981; Siddiqui et al. 1985; Green and Hadgraft 1987; Oakely and Swarbrick 1987) or ion-exchange (Siddiqui et al.1985, 1987). Thus, the ionisation state of a potential penetrant, in the context of its pK_a and the vehicle pH, will significantly affect the permeability of a molecule into and across the skin (Woolfson and McCafferty 1993a, b; Woolfson et al. 1998; Moss et al. 2006). Thus, the different aqueous solubilities of ionised and unionised species will influence the overall rate of permeability as flux is the product of the permeability coefficient, k_p, and the effective drug concentration in its vehicle (Williams 2003). Adjustment of the pH will therefore alter the amounts of penetrant available in the unionised or ionised forms, consequently affecting concentration, solubility and ultimately the rate of penetration across the skin (Woolfson et al. 1998; Williams 2003).

Physiological Factors Affecting Percutaneous Absorption

Skin Condition

Skin barrier and skin permeability normally relates to the ingress of chemicals into and across intact, healthy skin, and the determination of skin "quality" is significant in experiments which determine percutaneous absorption (see Chap. 2). This infers that models of skin permeation are therefore models of permeation across intact skin and do not represent a barrier compromised by, for example, diseases such as eczema or psoriasis where the barrier function of the skin is compromised or, in extreme cases, absent. Even if the skin is not broken, irritation and mild trauma may reduce the barrier to absorption. Mechanical damage, such as cuts and abrasions, or chemical burns may decrease the barrier properties of the skin and increase the rate of absorption. For example, Barry (1975) demonstrated that soaking excised *stratum corneum* in chloroform/methanol mixtures dramatically increased skin permeability due to the delipidisation of the barrier layer. Where the skin barrier is disrupted, it has been shown that absorption of hydrophilic solutes increases significantly more than hydrophobic molecules (Flynn 1985).

Skin Hydration and Occlusion

An increase in skin hydration is widely associated with an increase in the rate of penetration of most molecules. The exact nature and magnitude of such changes have been attributed to the physicochemical nature of the penetrant and the specific mechanism by which excess hydration is induced. Imokawa et al. (1991) suggested that the *stratum corneum* lipids were of significance as they held water in the skin through the formation of lamellar structures within the *stratum corneum*. Wiedmann (1988) suggested that the effective diffusion coefficient across the *stratum corneum* increases with an increase in water content, as the water content of the *stratum corneum* heightens the dynamic motion of epidermal tissue. The skin barrier has been shown to decrease rapidly over a short space of time—Auriol et al. (1993) suggested that significant decreases in barrier function could be observed after as little as ten minutes hydration of the skin. Skin hydration may also be influenced by the relative humidity of the external environment; changes in relative humidity have been shown to increase hydration and elevate the rate of diffusion (Fritsch and Stoughton 1963).

The process of skin occlusion involves entrapment of water which would normally be lost to the surrounding environment. This results in a rise in temperature at the skin surface and increased hydration of the occluded skin site (Zhai and Maibach 2001). It is most commonly achieved by placement of a water-impervious dressing on the skin or by the application of a highly viscous formulation (such as an ointment) which exerts a similar effect (Edwardson et al. 1993; Treffel et al. 1992). Occlusion of the skin in most cases leads to an increase in permeability. Indeed, in some cases, such as the application of the local anaesthetic products EMLA® Cream or Ametop™ gel, the use of a dressing which is intrinsically occlusive is important in increasing the efficacy and clinical effectiveness of these formulations. The volatility of the vehicle in which the penetrant is applied, and the physical nature of the penetrant, can also influence permeation. Stinchcomb et al. (1999) and Taylor et al. (2002) suggested that an increase in permeation under occlusive conditions is not always observed.

Skin Age

The structure and appearance of skin changes significantly with age, but it is often unclear if such changes are as a result of inherent ageing or influenced by environmental factors, or a combination of both. At the lower extreme of age, the infant (usually under two years of age) has, compared to adult skin, a higher water content and the *stratum corneum* barrier function is not fully formed. This means that skin in children under the age of two years is usually more permeable than adult skin (Barrett and Rutter 1994). The potential for increased permeation in such children should also be considered in the context of metabolism and drug delivery per kilogram of body weight in the context of the surface area to volume ratio (Plunkett et al. 1992). For example, it has been shown that the absorption of topical steroids is greater in children than in adults (Christophers and Kligman 1964; Idson 1975).

At the other extreme of age, it has been shown that alterations in keratinisation and epidermal cell production lead to changes in the intercellular spaces and a decrease in moisture content of skin (Rougier et al. 1988). However, the effects of these findings are not readily decoupled from other factors, such as any environmental influence on skin permeability or changes to the underlying skin vasculature and blood flow. Indeed, Roy and Flynn (1990) suggested that age was not a factor in the skin permeation of fentanyl and sufentanil, and that age-related permeability effects may not uniformly apply to all penetrants. They also concluded that, once fully formed, the *stratum corneum* maintains its barrier function.

Site-to-Site Variation

Wide variations in absorption rates have been found across different skin sites in the same individual and between different individuals. The permeability rates of molecules can generally be related to the thickness of the skin at particular points on the body. Wester and Maibach (1999) reported that this regional variation in absorption did not relate to the thickness of the *stratum corneum* as areas with the same thickness of *stratum corneum* demonstrated different permeability and areas with different thicknesses of *stratum corneum* demonstrated similar permeability.

Despite the inherent biological variation of skin ensuring that the overall process of skin permeability is complex and multifactorial, generalised trends in the wider literature suggest that the following ranking may be given to body sites (Scheuplein 1965; Feldman and Maibach 1967; Marzulli 1969; Elias et al. 1981):

posterior aricular skin > scrotum > head and neck > abdomen > forearm > thigh > instep > heel > planter

One clinically relevant example of this was the Transderm Scop® patch. This was a transdermal patch containing scopolamine, which is a drug with a poor percutaneous permeability profile. Therefore, patients were advised that the patch should be placed behind the ear due to the thinness of the posterior auricular skin. Therefore, regional variations in skin permeability can influence the site of application of medicinal products (Wester and Maibach 1999).

Race

The issue of whether race influence affects percutaneous absorption is complicated by the paucity of studies in this area. Of the few studies carried out, Lotte et al. (1993) suggested that there are no substantial differences between the permeability of African, Asian or European skin. They further suggested that greater skin pigmentation presents a greater barrier to absorption which recovers after perturbation more rapidly than more lightly pigmented skin. Bearardesca et al. (1991) highlighted the significant differences in *stratum corneum* water content between different races. However, the limited amount of research carried out in this field, coupled with the inherent variation in skin permeability, makes it difficult to draw definite conclusions on this subject.

Skin Temperature

It is generally accepted that an increase in temperature will increase the rate of absorption and that a decrease in temperature may lower the rate of absorption by up to one order of magnitude (Fritsch and Stoughton 1963; van der Merwe et al. 1988; Woolfson and McCafferty 1993a, b). An increase in temperature will also affect blood flow and metabolism. Percutaneous penetration usually occurs within a narrow temperature range, although occlusion may lead to an increase in temperature (Williams 2003). However, Allenby et al. (1969) suggested that little change in the rate of absorption is seen when the temperature is raised to 60 °C. Above this temperature, irreversible changes occur in the *stratum corneum*, affecting the arrangement of its lipids and their barrier function.

As skin permeation is initially a process of diffusion, it is therefore temperature dependant. The diffusion constant of a penetrant may be expressed by the Stokes-Einstein equation, in which temperature is prominent:

$$D = \frac{kT}{(6\pi r \eta)} \tag{1.5}$$

where D represents the diffusional constant, k represents the Boltzmann constant, T is the absolute temperature, r represents the hydrodynamic radius of the diffusing drug molecule, and η represents viscosity.

Vehicle Effects

As discussed above, percutaneous penetration is a series of diffusion and partition steps from, and between, a number of compartments. These rates rely on the collective effects that the skin, penetrant and vehicle exert on the diffusion process. In pharmaceutical applications, the vehicle allows optimisation and control of release at a rate adequate to provide a sufficient therapeutic dose of drug—such principles of the influence of a vehicle or formulation on permeability can also be applied to a number of related fields. A number of formulations influence, and increase, skin permeability by altering in some manner the structure—and hence barrier integrity—of the *stratum corneum*. The thermodynamic activity in the vehicle is the main driving force for a chemical to diffuse from the vehicle and then progress into and through the skin surface. In addition, the physicochemical properties of the penetrant will also influence its rate of diffusion. To optimise permeability, the vehicle must therefore present the permeant in a manner that will facilitate its rapid and/or controlled release from the vehicle to the skin. The pH of a vehicle will, as described above, also affect the activity coefficient of weakly acidic and basic molecules (Woolfson et al. 1998). Further, vehicles may affect the skin by increasing hydration and occlusion. For example, waxes and ointments are commonly found to increase hydration and therefore permeability through occlusion. Aqueous vehicles may occlude the skin less than non-aqueous systems, but they may increase hydration at the site of application, potentially increasing permeability. Bronaugh and Franz (1986) highlighted the significance of formulation and solvent choice, as they demonstrated that the permeation of caffeine, benzoic acid and testosterone formulated in three vehicles (petroleum, ethylene glycol gel and an aqueous gel) through human skin was significantly different. Ethanol has been widely employed as a solvent or cosolvent to increase the flux of molecules through the skin (Shahi and Zatz 1978; Idson 1983; Berner et al. 1989).

No universal vehicle exists for percutaneous absorption, particularly as there are a range of significant endpoints which have different goals—pharmaceutical systems will aim to optimise absorption, whereas other formulations, such as those with cosmetic applications, aim to reduce absorption into physiologically active tissues. The formulation must therefore be designed with the suitable endpoint in mind and to consider not just specific efficacies but wider issues of toxicity.

References

Allenby AC, Creasey NH, Edginton JAG, Fletcher JA, Schock C (1969) Mechanism of action of accelerants on skin penetration. Brit J Dermatol 81S:47–55

Anderson RL, Cassidy JM (1973) Variations in physical dimensions and chemical composition of human *stratum corneum*. J Invest Dermatol 61:30–32

Auriol F, Vaillant L, Machet L, Diridollou S, Lorette G (1993) Effects of short time hydration on skin extensibility. Acta Dermato-Venerelogica 73:344–347

Barker N, Hadgraft J (1981) Facilitated percutaneous absorption: a model system. Int J Pharm 8:193–202

Barlow HB, Mallon JD (eds) (1982) The senses. The University Press, Cambridge

Barr M (1962) Percutaneous absorption. J Pharm Sci 51:395–409

Barrett DA, Rutter N (1994) Percutaneous lignocaine absorption in newborn infants. Arch Dis Child Fetal Neonatal Ed 71:122–124

Barry BW (1975) Medicaments for topical application—biopharmaceutics of dermatological preparations. Pharm J 215:322–325

Barry BW (1983) Dermatological formulations: percutaneous absorption. Marcel Dekker, New York

Barry BW (1987) Mode of action of penetration enhancers in human skin. J Cont Rel 6:85–97

Beardesca E, de Rigal J, Leveque JL, Maibach HI (1991) In vivo biophysical characterisation of skin physiological differences in races. Dermatologica 182:89–93

Berner B, Mazzenga GC, Otte JH, Steffens RJ, Juang RH, Ebert CD (1989) Ethanol: water mutually enhanced transdermal theraputic system III: skin permeation of ethanol and nitroglycerin. J Pharm Sci 78:402–427

Blank IH (1964) Penetration of low molecular wight alcohols into the skin I effect of concentration of alcohol and type of vehicle. J. Invest. Dermatol. 43:415–420

Bouwstra JA, Gooris GS, Dubbelaar FE, Weerheim AM, Ponec M (1998) pH, cholesterol sulfate, and fatty acids affect the stratum corneum lipid organization. J Investig Dermatol Symp Proc 3:69–74

Brodal A (1981) Neurological anatomy in relation to clinical medicine. Oxford University Press, London

Bronaugh RL, Congdon ER (1984) Percutaneous absorption of hair dyes: correlation with partition coefficients. J. Invest. Dermatol. 83:124–127

Bronaugh RL, Franz TJ (1986) Vehicle effects on percutaneous absorption: *in vivo* and in vitro comparisons with human skin. Br J Dermatol 115:1–11

Bronaugh RL, Maibach HI (1999) Percutaneous absorption, 3rd edn. Marcel Dekker, Inc. CRC Press, New York

Christophers E, Kligman AM (1964) Percutaneous absorption in aged skin. In: Montagna W, (ed) Advances in the biology of the skin. Permagon, New York, p 163
Crank J (1975) The mathematics of diffusion, 2nd edn. Clarendon Press, Oxford
Cross SE, Roberts MS (1993) Subcutaneous absorption kinetics of interferon and other solutes. J Pharm Pharmacol 45:606–609
Edwardson PAD, Walker M, Breheny C (1993) Quantitative FT-IR determination of skin hydration following occulsion with hydrocolloid containing adhesive dressings. Int J Pharm 91:51–57
Elias PM (1988) Structure and function of the *stratum corneum* permeability barrier. Drug Develop Res 13:97–105
Elias PM, Cooper ER, Korc A, Brown BE (1981) Percutaneous transport in relation to *stratum corneum* structure and lipid composition. J Invest Dermatol 76:297–301
Elliott HC (1969) Textbook of neuroanatomy. Lippincott, Philadelphia
Feldman RJ, Maibach HI (1967) Regional variation in percutaneous absorption of ^{14}C-cortisol in man. J Invest Dermatol 48:181–183
Flynn GL (1985) Mechanism of percutaneous absorption from physicochemical evidence. In: Bronaugh RI, Maibach HI (eds) Percutaneous penetration. Dekker, London
Fritsch WC, Stoughton RB (1963) The effect of temperature and humidity on the penetration of ^{14}C-acetylsalicyclic acid in excised human skin. J Invest Dermatol 41:307–311
Green PG, Hadgraft J (1987) Facilitated transfer of cationic drugs across a lipoidal membrane by oleic acid and lauric acid. Int J Pharm 37:251–255
Heuber F, Wepierre J, Schaefer H (1992) Role of transepidermal and transfollicular routes in percutaneous absorption of hydrocortisone and testosterone—*In vivo* study in the hairless rat. Skin Pharmacol 5:99–107
Heuber F, Besnard M, Schaefer H, Wepierre J (1994) Percutaneous absorption of estradiol and progesterone in normal and appendage-free skin of the hairless rat—lack of importance of nutritional blood flow. Skin Pharmacol 7:245–256
Higo N, Sato S, Irie T, Uekama K (1995) Percutaneous penetration and metabolism of salicylic acid derivatives across hairless mouse skin in diffusion cell *in vitro*. STP Pharma Sci 5:302–308
Howes D, Black JG (1976) Percutaneous absorption of triclocarban in rat and man. Toxicol 6:67–76
Idson B (1975) Percutaneous absorption. J Pharm Sci 64:901–924
Idson B (1983) Vehicle effects in percutaneous absorption. Drug Met Rev 14:207–222
Imokawa G, Kuno H, Kawai M (1991) *Stratum corneum* lipids act as a bound water modulator. J Invest Dermatol 96:845–851
Le VH, Lippold BC (1995) Influence of physicochemical properties of homologous esters of nicotinic acid on skin permeability and maximum flux. Int J Pharm 124:285–292
Lotte C, Wester RC, Rougier A, Maibach HI (1993) Racial differences in the *in vivo* percutaneous absorption of some organic compounds: a comparison between black, Caucasian and Asian subjects. Arch Dermatol Res 284:456–459
MacKensie IC, Linder JC (1973) An examination of cellular organization within the *stratum corneum* by a silver staining method. J Invest Dermatol 61:254–260
Maibach HI, Feldman RJ (1969) Effect of applied concentration on percutaneous absorption in man. J Invest Dermatol 52:382
Martin A, Swarbrick J, Cammarata A (1983) Physical Pharmacy, 3rd edn. Lea & Febinger, Philadelphia
Marzulli FN (1969) Barriers to skin penetration. J Invest Dermatol 39:387–393
Matoltsy AG (1976) Keratinisation. J Invest Dermatol 67:20–25
Michaelis AS, Chandrasekaran SK, Shaw JE (1975) Drug permeation through human skin: theory and *in vitro* experimental measurement. AIChE. 21:985–996
Michel S, Schmidt R, Shroot B, Reichert U (1988) Morphological and biochemical characterization of the cornified envelopes from human epidermal keratinocytes of different origin. J Invest Dermatol 91:11–15

Mitragotri S, Anissimov YG, Bunge AL, Frasch HF, Guy RH, Hadgraft J, Kasting GB, Lane ME, Roberts MS (2011) Mathematical models of skin permeability: an overview. Int J Pharm 418:115–129

Montagna W (1964) The skin of the domestic pig. J Invest Dermatol 42:11–21

Moss GP, Dearden JC, Patel H, Cronin MTD (2002) Quantitative structure-permeability relationships (QSPRs) for percutaneous absorption. Toxicol In Vitro 16:299–317

Moss GP, Woolfson AD, Gullick DR, McCafferty DF (2006) Mechanical characterisation and drug permeation properties of tetracaine-loaded bioadhesive films for percutaneous local anaesthesia. Drug Dev Ind Pharm 32, 163–174

Moss GP, Wilkinson SC, Sun Y (2012) Mathematical modelling of percutaneous absorption. Curr Opin Coll Interf Sci 17:166–172

Nakaue HS, Buhler DR (1976) Percutaneous absorption of hexachlorophene in the rat. Toxicol Appl Pharmacol 35:381–391

Norlen L (2006) Stratum corneum keratin structure, function and formation—a comprehensive review. Int J Cosmet Sci 28:397–425

Norlen L (2007) Nanostructure of the stratum corneum extracellular lipid matrix as observed by cryo-electron microscopy of vitreous skin sections. Int J Cosmet Sci 29:335–352

Oakely DM, Swarbrick J (1987) Effects of ionization on the percutaneous absorption of drugs: partitioning of nicotine into organic liquids and the hydrated *stratum corneum*. J Pharm Sci 76:866–871

Ostrenga J, Steinmetz C, Poulsen B, Yett S (1971a) Significance of vehicle composition II: prediction of optimal vehicle composition. J Pharm Sci 60:1180–1183

Ostrenga J, Steinmetz C, Poulsen B (1971b) Significance of vehicle composition I: relationship between topical vehicle composition, skin penetrability and clinical efficacy. J Pharm Sci 60:1175–1179

Parry GE, Bunge AL, Silcox GD, Pershing LK, Pershing DW (1990) Percutaneous absorption of benzoic acid across human skin. 1. *In vitro* experiments and mathematical modelling. Pharm Res 7:230–236

Pilgram GS, Engelsma-van Pelt AM, Bouwstra JA, Koerten HK (1999) Electron diffraction provides new information on human stratum corneum lipid organization studied in relation to depth and temperature. J Invest Dermatol 113:403–409

Pilgram GS, van der Meulen J, Gooris GS, Koerten HK, Bouwstra JA (2001) The influence of two azones and sebaceous lipids on the lateral organization of lipids isolated from human stratum corneum. Biochim Biophys Acta 511:244–254

Plunkett LM, Turnbull D, Rodricks JV (1992) Differences between adults and children affecting exposure assessment. In: Guzelian PS, Henry CJ, Olin SS (eds) Similarities and differences between children and adults, implications for risk assessment. ILSI Press, Washington, p 79–94

Rawlings AV (2003) Trends in stratum corneum research and the management of dry skin conditions. Int J Cosmet Sci 25:63–95

Rawlings AV (2010) Recent advances in skin 'barrier' research. J Pharm Pharmacol 62:671–677

Rougier A, Lotte C, Corcuff P, Maibach HI (1988) Relationship between skin permeability and corneocyte size according to anatomical age, site and sex in man. J Soc Cosmet Chem 39:15–26

Roy SD, Flynn GL (1990) Transdermal delivery of narcotic analgesics—pH, anatomical and subject influences on cutaneous permeability of fentanyl and sufentanil. Pharm Res 7:842–847

Scheuplein RJ (1965) Mechanism of percutaneous absorption I routes of penetration and the influence of solubility. J Invest Dermatol 45:334–346

Scheuplein RJ (1967) Mechanism of percutaneous absorption I transient diffusion and the relative importance of various routes of skin penetration. J Invest Dermatol 48:79–88

Scheuplein RJ, Blank IH (1971) Permeability of the skin. Physiol Rev 51:702–747

Scheuplein RJ, Blank IH, Brauner GI, MacFarlane DJ (1969) Percutaneous absorption of steroids. J Invest Dermatol 52:63–70

Shahi V, Zatz JL (1978) Effect of formulation factors on penetration of hydrocortisone through mouse skin. J Pharm Sci 67:789–792

Shore PA, Brodie BB, Hogben CAM (1957) The gastric secretion of drugs: a pH partition hypothesis. J Pharmacol Exp Ther 119:361–369
Siddiqui O, Roberts MS, Polack AE (1985) Topical absorption of methotrexate: role of dermal transport. Int J Pharm 27:193–203
Siddiqui O, Sun Y, Liu JC, Chien YW et al (1987) Facilitated transdermal transport of insulin. J Pharm Sci 76:341–345
Sinclair D (1981) The mechanisms of cutaneous sensations, 2nd edn. Oxford University Press, London
Sinko PJ (2005) Martin's Physical Pharmacy and Pharmaceutical Sciences. 5th revised edn. Lippincott, Williams and Wilkins, Baltimore
Stinchomb AL, Pirot F, Touraille GD (1999) Chemical uptake into human *stratum corneum in vivo* from volatile and non-volatile solvents. Pharm Res 16:1288–1293
Sved S, McClean WM, McGilvernay IJ (1981) Influence of the method of application on the pharmacokinetics of nitroglycerin from ointments in humans. J Pharm Sci 70:1368–1369
Swarbrick J, Lee G, Brom J, Gensmantel NP (1984) Drug permeation through the human skin II: permeability of ionized compounds. J Pharm Sci 73:1352–1355
Taylor LJ, Lee RS, Long M, Rawlings AV, Tubek J, Whitehead L, Moss GP (2002) Effect of occlusion on the percutaneous penetration of linoleic acid and glycerol. Int J Pharm 249:157–164
Treffel P, Muret P, Muret-D'Aniello P, Coumes-Marquet S, Agache P (1992) Effect of occlusion on in vitro percutaneous absorption of two compounds with different physicochemical properties. Skin Pharmacol 5:108–113
Treherne JE (1956) Premeability of skin to some non-electrolytes. J Physiol 133:171–180
van der Merwe E, Ackermann C, van Wyk CJ (1988) Factors affecting the permeability of urea and water through nude mouse skin *in vitro* I temperature and time of hydration. Int J Pharm 44:71–74
Weddell G (1941) The pattern of cutaneous innerveration in relation to cutaneous sensibility. J Anat 75:346–367
Wester RC, Maibach HI (1999) Regional variation in percutaneous absorption. In: Bronaugh RL, Maibach HI (eds) Percutaneous absorption; drugs—cosmetics—mechanisms—methodology, 3rd edn. Marcel Dekker, New York p 107–116
Wester RC, Noonan PK (1980) Relevance of animal models for percutaneous absorption. Int J Pharm 7:99–110
Wester RC, Noonan PK, Maibach HI (1977) Frequency of application on the percutaneous absorption of hydrocortisone. Arch Dermatol Res 113:620–622
Wiedmann TS (1988) Influence of hydration on epdiermal tissue. J Pharm Sci 77:1037–1041
Wilkes GL, Brown IA, Wildnauer RH (1973) The biomechanicalproperties of skin. CRC Crit Rev Bioeng 1:453–495
Williams AC (2003) Transdermal and topical drug delivery. The Pharmaceutical Press, London
Wilson JS, Holland LM (1982) The effect of application frequency on epidermal carcinogenesis assays. Toxicol 24:45–54
Woolfson AD, McCafferty DF (1993a) Percutaneous local anaesthesia: drug release characteristics of the amethocaine phase-change system. Int J Pharm 94:75–80
Woolfson AD, McCafferty DF (1993b) Percutaneous local anaesthesia. Ellis Horwood, London
Woolfson AD, McCafferty DF, Moss GP (1998) Development and characterisation of a moisture-activated bioadhesive drug delivery system for percutaneous local anaesthesia. Int J Pharm 169:83–94
Zhai HB, Maibach HI (2001) Effects of skin occlusion on percutaneous absorption: an overview. Skin Pharmacol Appl Skin Physiol 2001(14):1–10

Chapter 2
Methods for the Measurement of Percutaneous Absorption

In general, the models of skin absorption discussed herein are generated from data published in the literature. These laboratory-based methods may use a variety of methods which could potentially affect the results of the experiments, particularly when used in subsequently developing a mathematical model. Therefore, the aim of this chapter is to discuss the different experimental protocols that are commonly used by researchers in percutaneous absorption in the context of the potential impact such methods of producing a model's inputs might exert upon its output.

Introduction

The vast majority of mathematical estimates of percutaneous absorption use, as their primary input, information on the rate of passage, or permeability, of a chemical across the skin. This is usually the permeability coefficient, k_p, or the more infrequently used (in the context of model development) maximum steady-state flux, J_{max}. In addition, a number of the physicochemical descriptors also modelled are measured experimentally, including measures of lipophilicity (commonly referred to as the octanol–water partition coefficient, log P) and melting point.

While different experiments will output the same general information—k_p or J_{max}—they may derive this information using different experimental protocols. Thus, the nature of the experiment and how it influences our understanding of permeability, not just its application to modelling, is a significant issue. For example, while there may be good reasons for using a range of experimental protocols to determine the permeability of particular penetrants, it is important to understand how this may apply itself to the subsequent—and, we should remember, the very separate—exercise of developing a mathematical model with this data.

The aim of this chapter is to discuss the different experimental protocols that are commonly used by researchers in percutaneous absorption, often to answer very specific experimental questions, contextualising our understanding of where the data used to develop models comes from and how different methods of generating the data might influence the output of models thus derived. It should be noted that

G.P. Moss et al., *Predictive Methods in Percutaneous Absorption*,
DOI 10.1007/978-3-662-47371-9_2

this chapter is not a full description of this field; rather, it highlights the aspects of experimental design that are most relevant for the development of mathematical models. For a comprehensive discussion of this subject, the reader is directed to Bronaugh and Maibach's (1999) or Williams' (2003) excellent texts.

In Vivo and In Vitro Methods: Overview

It should be first commented that the vast majority of mathematical models for percutaneous absorption used data from, and therefore most closely reflect, in vitro laboratory experiments. These are physical experiments that use membranes, which are either mammalian or synthetic in nature, across which the permeation of a chemical is measured experimentally. Such experiments are widely carried out and are an area of substantial interest across a range of industries. They have been used to measure the percutaneous absorption of pharmaceuticals, materials in cosmetic formulations, for toxicology studies and for estimation of risk assessment and occupational exposure of materials used in a variety of industrial applications.

In vitro methods are commonly used prior to in vivo experiments and in some cases (such as for the assessment of new chemical entities) are solely used to provide an indication of potential toxicity prior to any human exposure. Consequently, in vitro models are widely and commonly employed to assess the risks and hazards associated with exposure of human skin to exogenous chemicals.

Classically, in vivo studies have been conducted and provided valuable information on the mechanism of percutaneous absorption. However, these studies were generally non-invasive in that they measured a response in the skin, such as vasodilatation or skin blanching, rather than taking blood samples or punch biopsies of the skin for subsequent analysis. Despite their advantages, such methods are clearly limited in their applicability to other chemicals, particularly those that do not result in a non-invasively measurable physiological change. In addition, the non-invasive monitoring of certain topically applied chemicals, such as cosmetic formulations, may be measured in terms of efficacy by a range of biophysical methods, but such methods generally (with the exception of, for example, patch testing) do not provide any indication of cutaneous toxicity.

The in vivo estimation of percutaneous absorption may be considered appropriate if an established material (such as the drug ibuprofen) is used, and its absorption, distribution, metabolism and elimination are estimated by the analysis of bodily fluids. This, however, is extremely difficult to do for a wide range of potential penetrants—not just for the toxicological reasons mentioned above—but for logistical reasons, particularly the consistent availability of volunteers. It is also potentially unethical, should novel materials or techniques be investigated, such as the use of chemical or physical methods of enhancing absorption (i.e. formulation-based approaches or the use of electrical currents—iontophoresis—to facilitate absorption). In vivo experiments can provide realistic information on the amount of a topically applied chemical that is absorbed into and across the skin and which becomes bioavailable. However, in the context of the mathematical modelling of percutaneous absorption, the vast majority of models are based on in vitro experiments using excised human skin, as the paucity of in vivo data, and lack of consistent endpoints (i.e. the measurement of a penetrant in a body compartment or the use of a non-invasive clinical response) means that there is insufficient data available in the literature from which a valid model can be constructed. Thus, while not lessening the overall significance in the wider field of percutaneous absorption of in vivo testing, the main focus of this chapter will be on in vitro methods for the measurement of percutaneous absorption. Clearly, in vitro methods are informed by, and attempt to replicate, in vivo methods and it is in that context that the biorelevance of in vitro testing should be considered.

In Vitro Experimental Methods

Preamble

In vitro methods for the characterisation of percutaneous absorption, while ultimately delivering the same outcome, are many and varied in the details of their methods. Selection of the diffusion membrane, type of cell (i.e. the use of either "static" or "flow-through" cell designs, described below), nature of the experiment (e.g. duration, occlusion) and the composition of the phases that sit either side of the diffusion membrane are some of the key parameters that add to the diversity of acceptable experimental protocols from which the data to construct mathematical models is abstracted. The main issues in the experimental design for the measurement of percutaneous absorption are discussed below.

Membrane Selection

In vitro methodologies for the measurement of percutaneous absorption are widely established and validated (Franz 1975; Maibach 1975; Mali 1978; Bronaugh et al. 1982; Barry 1983; Bronaugh and Maibach 1985a, b; Friend 1992). They are used commonly in a variety of research fields as described above. Human skin from various sources, including cosmetic surgery and amputations, has been widely used for the in vitro assessment of percutaneous penetration (Franz 1975; Bronaugh et al. 1982; Barry 1983; Friend 1992). The use of various animal skins is also a commonly accepted constituent of in vitro percutaneous penetration studies. Skin from a wide range of species, including pigs, rats, guinea pigs, monkeys and snakes, among others, has been suggested as a suitable replacement for human skin (Bartek et al. 1972; Marzulli and Maibach 1975; Wester and Maibach 1976; Chow et al. 1978; Wester and Noonan 1980; Itoh et al. 1990; Roberts and Mueller 1990; Sato et al. 1991; Lin et al. 1992; Harada et al. 1993). Generally, skin from the pig and the rat has found the most widespread use, with the former in particular offering similar barriers to diffusion for the penetration through human skin of a wide range of molecules. Rat or mouse skin may be much more (up to 10 times) permeable than human skin, while pigskin has been claimed to be a better surrogate (Bartek et al. 1972; Chow et al. 1978; Wester and Noonan 1980; Roberts and Mueller 1990; Sato et al. 1991; Lin et al. 1992; Harada et al. 1993). However, rodent skin is still widely used as an in vitro membrane, possibly due to the use of such species more broadly in pharmacological research.

Several researchers have developed artificial skin equivalents, often known as living skin equivalents (LSEs) in an attempt to address some of the issues associated with using animal tissue in place of human skin (such as the lack of similarity in diffusional characteristics or complexity compared to human skin, and the *stratum corneum* in particular). LSEs have been used with some success in skin grafting and in the surgical treatment of burns (Young et al. 1998; Berger et al. 2000; Kremer et al. 2000; Machens et al. 2000; Mizunuma et al. 2000; Yang et al. 2000). Such materials aim to replicate the hydrophilic and hydrophobic balance of human *stratum corneum*, as well as the manifestation of its barrier function in, for example, the control of transepidermal water loss (TEWL) and control of bacterial ingress to the deeper epidermal and dermal tissues.

LSEs have also been used to assess percutaneous absorption. They generally consist of skin membranes which may include reconstituted epidermal cells that have been grown in tissue culture. They were proposed as an alternative to animal skin for in vitro percutaneous permeation studies but have to date failed to gain widespread acceptance. This is due to the reproducibility, cost (particularly when compared to animal tissue and where a large number of replicates of an experiment are required), their lack of robustness compared to human or animal skin (i.e. particularly when a formulation has to be directly applied to the skin, such as a semi-solid in a manner consistent with its clinical or consumer use) and their ability to replicate these tissues in terms of permeability and other physical properties. Several researchers have demonstrated that LSEs can have similar diffusional characteristics to mammalian skin but that they generally overestimate the rate of permeation across the membrane (Pelle et al. 1993; Hager et al. 1994; Horiguchi et al. 1997; Nemecek and Dayan 1999; Ramsamooj et al. 1998; Wang et al. 2000).

Artificial membranes have been used when human or animal skin is difficult to obtain, or where a large number of experiments are to be carried out, particularly with regard to preformulation screening experiments. The most widely used artificial membranes are polydimethylsiloxane (PDMS) and cellulose acetate (porous dialysis tubing) (Kurosaki et al. 1991; Megrab et al. 1995a, b; Stott et al. 2001; van Hal et al. 1996; Esposito et al. 1998; Woolfson et al. 1998; Minghetti et al. 1999). However, these membranes have often been shown to overestimate significantly the flux across skin and their use is significantly limited. For example, Moss et al. (2006) compared the permeability of a series of prodrugs across pigskin and PDMS membranes in vitro. They demonstrated a reasonable relationship for hydrophilic molecules, whereas an increase in hydrophobicity resulted in a significant difference in permeability, with the PDMS showing significant overestimation of permeability compared to pigskin.

However, while studies using animal skin are experimentally sound and widely used, it should be noted that very few mathematical models of percutaneous absorption are based on animal tissue. This may be due to the relatively early publication of the seminal 1992 paper by Potts and Guy which developed quantitative structure–permeability relationships using human skin—thereafter, the development of a mathematical model based on skin from other species would not have the same significance and would almost appear to be a retrograde or rather redundant step, despite the opportunity to glean, in a comparative sense, important mechanistic information on the relative value of using non-human skin, as has been achieved by comparing mathematical models of human skin to an artificial membrane, polydimethylsiloxane, which also finds widespread use as a substitute for animal tissue in such experiments (Potts and Guy 1992; Geinoz et al. 2002; Moss et al. 2006).

Hence, the use of animal skin, artificial membranes or skin equivalents in place of human skin is limited, but does highlight the issue, particularly in the context of its extensive use in the wider field of percutaneous absorption, the importance of mathematical models in relation to elucidating the mechanism of action of permeants.

Such a discussion may infer that the use of human tissue is optimal for in vitro experiments. However, it is not without its problems. Human skin is usually obtained from skin banks or from tissue donated by the patient as a result of surgery—generally, this can range from cosmetic procedures including face lifts, "tummy tucks" and breast reduction or from medical procedures including amputations. The experimenter therefore has little control over the handling and quality of the skin obtained. For example, surgical procedures and protocols used to remove skin may include the use of alcohol-based disinfectants; while this is clearly essential for the surgical procedure and is a central part of infection control policies, it may affect the permeability in subsequent experiments as the alcohol has the potential to remove *stratum corneum* lipids and potentially affect its barrier properties. Further, the skin may be frozen and stored prior to dispatch and use in experiments, and this may result in damage to, or degradation of, the membrane.

Further, variability of skin was discussed in Chapter One and may be due to gender, age, body site and a number of other factors. Tissue from different body sites and from different patients may, not unreasonably, exhibit different properties, including permeability. In terms of experimental use, such variation is generally minimised by attempting to use tissue that is as similar as possible (i.e. from the same region, or from the same patient, or from a small patient population) or by normalising against a known standard. While there are different implications for the type of experiment being conducted (i.e. a comparative formulation study might require different standards than a toxicological study), all such data from a range of experiments have been collated into data sets and used to develop mathematical models of skin permeability, thus introducing a possible source of variance—or potentially even error—into the models. While this is a significant issue, it should ultimately be considered in the context of tissue availability as discussed above, and any modellers should be aware of such issues, particularly the history and nature of the tissue, and consider their output accordingly.

The skin used in percutaneous absorption experiments has to be prepared before use. This involves several stages. Firstly, once the skin has been harvested surgically, it is used either fresh or frozen. Keeping human skin frozen at −20 °C has been shown to have no detrimental effect on its permeability for up to 455 days (Harrison et al. 1984). Consequently, with regard to availability of supply and convenience, skin for in vitro permeation studies is frequently used from frozen, being defrosted prior to use. Skin samples are normally defrosted at ambient temperature and gently patted dry before use.

The most rudimentary preparation immediately prior to an experiment involves removing the surface hairs, or clipping them to a length where they will not interfere with the experiment by, for example, compromising the seal between the skin surface and the compartment directly above it (ensuring that this is done in a manner that will not compromise the skin barrier), and removing subcutaneous fat normally by the careful application of dissection techniques. Such processing of the skin may also be conducted prior to freezing the skin samples. However, there are instances when specific layers of the skin—most normally, the epidermis or the *stratum corneum*—are isolated from the rest of the tissue and used rather than whole skin.

Next, if the experimental protocol determines that the skin has to be separated in some manner, several methods are available to facilitate this. For example, the use of blistering, either by the application of chemicals or suction, can be used although it may damage the skin and cause issues with membrane integrity. Skin may be processed to the required thickness by the use of a dermatome. There are a number of methods commonly employed for this, and one good example of the difficulties encountered is described by Williams (2003). He reports a technique that involves freezing the predominately aqueous dermal side of the excised skin to a steel plate. This fixes the skin to the steel surface, and once it has thawed slightly, it can be dermatomed to the required thickness.

Alternatively, chemicals may be used to separate either the *stratum corneum* or the entire epidermis from the underlying tissue. Elias et al. (1974) used staphylococcal epidermolytic toxin and ammonia fumes to separate the epidermis from the dermis. A more widely used method involves the application of heat to separate the epidermis from the dermis (Kligman and Christophers 1963). The skin is placed in water at 60 °C for a short duration (usually 45 s to several minutes is recommended), removed, patted dry and the epidermis removed, usually by pinning the skin onto a cork board and peeling the epidermis off.

Use of isolated *stratum corneum* involves separation from the rest of the skin by enzymatic methods, such as the use of a trypsin solution, where the skin sample is placed, *stratum corneum* side upwards, onto a solution of trypsin. Tissue from the viable epidermis and the dermis is removed by this process, and any residue remaining on the skin surface is subsequently gently removed, and the resulting membrane is dried and stored. Such methods are rarely used as the isolated *stratum corneum* is very thin and fragile—also, in a diffusion cell, the fragile *stratum corneum* may also require a support to maintain integrity of the diffusion cell. Removal of the *stratum corneum* from the underlying viable epidermis may result in tears to the barrier, and therefore, barrier integrity checks need to be carried out before and after permeation experiments using this tissue.

It is more common to prepare the skin by using epidermal tissue which has been separated from the lower layers of the skin by heat. As the *stratum corneum* is the main barrier to percutaneous absorption, the tissues of the viable epidermis, which are predominately aqueous in nature, essentially provide little more than physical and mechanical support and improved flexibility compared to isolated *stratum corneum*. Removal of the dermis from the experimental tissue has little effect in the process of percutaneous absorption as the main clearance of materials absorbed into the skin occurs at the top of the dermis, at its junction with the epidermis, where the majority of the skin's vasculature that clears such permeants resides. Conversely, as the dermis is also a predominately aqueous tissue which, in the context of an in vitro percutaneous absorption study, provides physical and mechanical support rather than a diffusing stratum, it is often included in experiments as its presence will have little effect on the overall process of absorption. However, it may in practice mean that the skin is excessively thick and therefore cumbersome in the diffusion apparatus, and it is often removed, wholly or partially, by the use of a dermatome which results in the use of a consistent thickness of skin.

Williams (2003) elaborates on such aspects of experimental design by considering them in the context of the penetrant's solubility. If the penetrant is hydrophilic (defined as having an octanol–water partition coefficient below 3), the main barrier to diffusion across the skin will reside in the *stratum corneum* and the absence of dermal tissue is not an issue. However, in the case of lipophilic penetrants (defined as having an octanol–water partition coefficient above 3), the lower, predominately aqueous layers of the skin may become significant as they will provide a substantial challenge to the permeation of such materials and, at the very least, an epidermal membrane, rather than dermatomed skin, would be the preferred membrane.

Integrity Testing

The substantial amount of processing of the skin undergoes prior to its use in a diffusion experiment means that it is susceptible to damage, particularly to the fine *stratum corneum* barrier. Thus, assessment of skin integrity is then carried out to determine the viability of the tissue. This usually involves the measurement of a particular property of the skin, or the passage of a "marker" compound through the skin prior to the experiment. Water naturally evaporates from the body and across the skin in a process called transepidermal water loss (TEWL). The rate of evaporation can be measured by sensitive equipment (i.e. typical value 0.5 $\mu L/cm^2/h$; Imhof et al. 2009) and the rates compared to known benchmarks, or the standard deviation can be determined for a batch of skin being used in a particular experiment, or set of experiments, and those samples whose TEWL falls outside the chosen range can be discarded as being atypical. Measurement of TEWL is non-invasive and does not alter the integrity of the skin by the direct application of chemicals prior to the experiment. TEWL is sensitive to the environmental factors, particularly relative humidity. Similarly, the electrical resistance of skin can be measured by placement of electrodes either side of the skin tissue. As with TEWL measurements, determination of electrical resistance is non-invasive and atypical values, depending on the criteria adopted by researchers, can lead to the rejection of particular samples from the experiment. Another option is to use a chemical to determine the integrity of the skin barrier. Tritiated compounds (^{3}H), including tritiated water or radiolabelled sugars such as glucose, have been used by several researchers to ensure the integrity of the skin barrier for a period of time (usually not more than 6–8 h) prior to an experiment (Bronaugh et al. 1986; Harrison et al. 1984). Skin is usually accepted for use if the permeability coefficient (cm/h) of the penetrant is within a certain range—often, while there is a general consensus as to what constitutes "acceptable", different researchers use comparable but often subjective measurements to remove poor atypical tissue from an experiment. ^{3}H and other radiolabelled compounds do not find widespread use in percutaneous absorption experiments due to the logistics and costs associated with such experiments. With the skin prepared and its integrity checked, the apparatus can now be assembled.

Selection of the Diffusion Cell Apparatus

Fundamentally, the aim of an in vitro percutaneous absorption study is to determine the amount of permeant that passes into and across the skin. This involves the use of a diffusion chamber in which the membrane (normally human or animal skin, or an artificial membrane) separates the two compartments—the donor compartment is where the formulation containing the permeant of interest is introduced at the start of the experiment; and the receptor compartment is the chamber into which the permeant of interest may diffuse, following passage into and across the membrane. In addition to measuring the amount, and rate, of permeant that passes across the membrane and into the receptor phase, the experiment also presents the experimenter with the opportunity to determine how much of the material of interest has passed into the skin and has remained there at the end of the experiment. This is usually achieved by using adhesive tape to remove the *stratum corneum* and by digesting the remaining tissue using acid/solvent mixtures. These samples can then be prepared for quantitative analysis.

Thus, the experiment is essentially a passive diffusion process that is governed by the diffusion gradient across the membrane as well as the experimental protocol and the physicochemical properties of the permeant. Despite various designs, some more complex than others (Fig. 2.1), the fundamentals of the process are very similar—passage of the permeant of interest from the donor chamber to the receptor chamber while maintaining a viable diffusion gradient and avoiding equilibrium

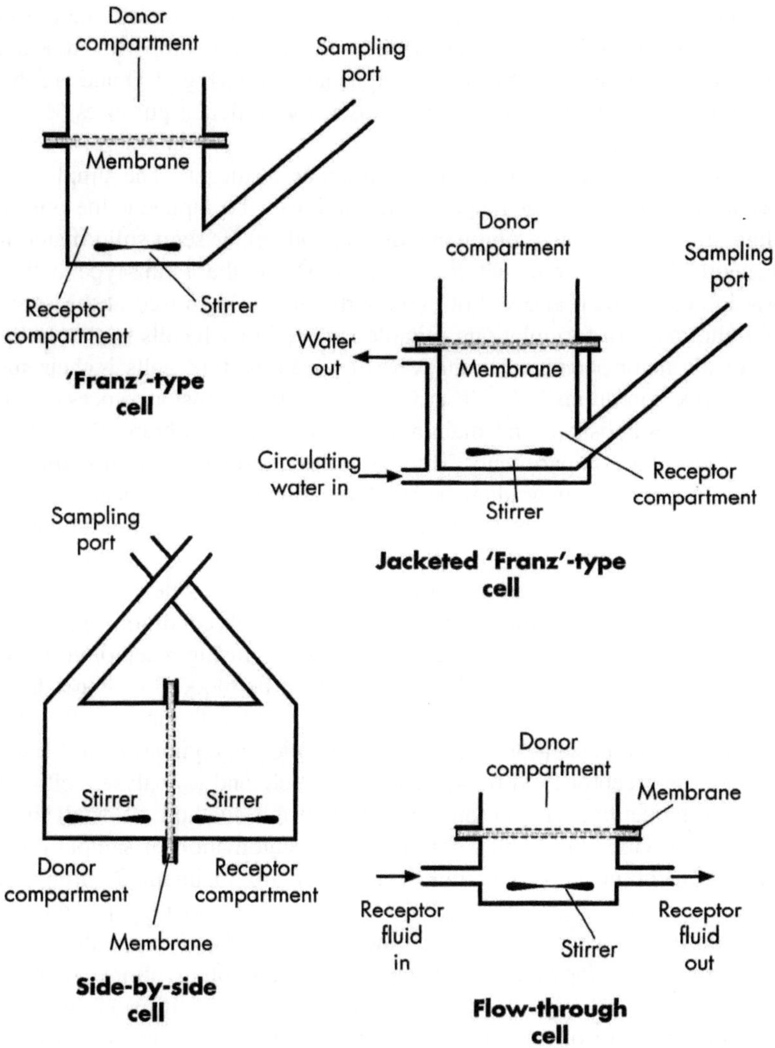

Fig. 2.1 Examples of commonly used diffusion cells (© Williams (2003), used with permission)

between both compartments. The receptor compartment normally has a sampling port—usually a sidearm—which is occluded during the experiment to stop evaporative loss and from which aliquots of the receptor media may be removed (and replaced with fresh receptor fluid) for analysis.

There are two main types of diffusion cells: static and flow-through cells. The vast majority of in vitro percutaneous absorption experiments are conducted using upright static glass diffusion cells, known as Franz or Franz-type cells (Franz 1975). These involve the partition of the permeant into a "static" receptor compartment of a fixed volume that is maintained at a controlled temperature and is continually stirred throughout the experiment to ensure complete mixing and to avoid the presence of any "dead" zones or diffusion gradients within the receptor compartment. Franz-type diffusion cells are found in a range of designs and sizes (Fig. 2.1) and are therefore very flexible, with particular cells being chosen for particular purposes (e.g. when a different surface area might be required, or when the receptor compartment may need to be varied based on the solubility of a particular permeant). The simple design of the upright cells allows a wide range of formulations to be applied to the skin surface, including solutions, creams, ointments and various other semi-solid materials of a pharmaceutical or cosmetic nature. A variation on the Franz-type cells is the side-by-side cell, which allows both compartments to be stirred at the same time, although the range of formulations suitable for use in such cells is clearly limited.

One of the main perceived limitations of the Franz-type cells is their susceptibility to "sink conditions". As described above, the diffusion process requires a gradient to be established, and maintained, across the membrane that divides the donor and receptor compartments. Thus, in a static diffusion cell, it is important to maintain a concentration gradient such that, normally, the concentration of the permeant of interest is not greater than 10 % of its saturated solution in the receptor compartment (Martin et al. 1983; Anissimov and Roberts 1999; Roberts and Cross 1999; Roberts et al. 1999). Flow-through diffusion cells, often called "Bronaugh cells", offer an alternative approach to the issue of sink conditions and attempt to mimic in vivo conditions by utilising a constantly perfusing receptor compartment, flowing usually at 1–2 ml/h, which aims to mimic the blood flow beneath the skin (Bronaugh et al. 1999a, b). This is normally achieved with the use of a peristaltic pump which is connected to tubing that supplies a compartment beneath the membrane. The receptor fluid flows into a receptacle and is analysed, either offline or, more infrequently, via an online flow-through method of analysis. Either arrangement provides the facility for substantial automation of sample collection. This arrangement ensures that sink conditions are maintained throughout the experiment and that the receptor phase does not have to be replaced should a penetrant be rapidly absorbed. While flow-through cells are more flexible (i.e. in terms of automation), they are more complex and substantially more expensive than Franz cells. Consequently, the use of Franz-type static diffusion cells is more common than the use of flow-through cells. In either case, most experiments use between 6 and 24 cells per experiment in order to ensure reproducibility.

As mentioned above, the cell design can influence the formulation type that can be applied to it. For example, the side-by-side cells, while having the advantage of even stirring and mixing in both donor and receptor compartments, are limited in the type of formulation that can be applied—pharmaceutical dosage forms such as creams, gels, ointments and patches are very difficult to apply in this manner. Thus, the cell design and the protocol used can all affect the outcome—the measurement of percutaneous absorption and the cell design should be considered when considering the uniformity of data that are to be used in the construction of mathematical models of skin absorption.

For example, the manner in which the product is applied to the skin surface is important to consider in model construction, as each individual experiment may require specific amounts, or even frequencies of application, or they may require that the donor phase is occluded, or not occluded—the experimental protocol used is dependent on the nature of the penetrant and the context of the experiment. An experiment might wish to examine the occupational hazard of a chemical being exposed to the skin—in, for example, the context of crop spraying of a pesticide. Therefore, an experimental protocol might use a non-occluded experiment (to mimic the potential of the solvent, the pesticide is delivered in evaporating vehicle) and a fixed, or finite, dose which would represent the amount commonly sprayed and therefore the amount that the skin (per unit area) might be exposed to. Equally, an experiment may wish to determine the "worst-case scenario" for the application of a new chemical to the skin, and, in such circumstances, an occluded experiment may be favoured as this generally increases the permeation of a topically applied chemical (Zhai and Maibach 2001). Such a protocol might also use an excessive amount of chemical—an infinite dose—that aims to maintain the maximum flux of this chemical across the skin, in order to determine permeation in the "worst case". Such an arrangement also finds use when examining the fundamental behaviour of a material and is used to establish the steady-state rate of permeation across the skin. Infinite doses are also used when considering the effects of formulation ingredients that alter the barrier function of skin (penetration enhancers) on the penetration of a chemical (Williams 2003).

The application of an infinite dose to the surface of the skin means that there is little or no significant change to its concentration (in the donor phase) or thermodynamic activity throughout the experiment. However, once permeation commences, the donor phase becomes depleted and this scenario is no longer valid. In a practical sense if the concentration in the receptor does not exceed 10 % of the saturated solubility, an infinite dose can be assumed. Further, supersaturated systems have been employed (e.g. Raghavan et al. 2001; Dias et al. 2003; Raghavan et al. 2003) to resolve this issue; the donor phase is composed of a saturated solution of the penetrant of interest, into which is also added an additional amount of the penetrant, which will be present as a solid. As diffusion into and across the skin commences, the donor phase can maintain the saturated solution should the conditions, including the rate of solubility, permit it to do so (and, therefore, maintain in theory the same concentration and thermodynamic activity) throughout the duration of the experiment.

Conversely, in a finite-dose experiment, a fixed amount of a material is applied to the skin surface and it is expected that it will deplete during the course of the experiment. This may additionally be facilitated by whether or not the donor phase is covered (occluded) during the experiment, as a non-occluded system may result in the evaporation of any solvent which may alter, both positively and negatively, the permeation of the material. Such experiments are very useful in that they allow an estimation of permeation that is more realistic and which more accurately represents the "in-use" performance of a product.

Temperature

The temperature at which an in vitro percutaneous penetration experiment is conducted is normally 37 °C, maintained by the use of circulating water, as discussed above. This results in a temperature of 32 °C at the skin surface. It is important to maintain this temperature throughout the experiment as skin permeability may vary significantly with even small changes in temperature (Chilcott et al. 2005). However, data sets collated for several studies (i.e. Moss et al. 2009; Lam et al. 2010) show that, while the vast majority of experiments that constitute those data sets have been conducted at either 32 or 37 °C, a significant number of experiments were conducted at temperatures of 31–32, 25–31 and 22–30 °C. The implications of this are discussed in later chapters.

Formulation and Solubility Factors

The nature of any formulation applied to the skin must be considered when using the information from an in vitro diffusion cell experiment to construct a mathematical model of permeation. If a chemical applied to the skin, whether it is the permeant of interest or a material formulated to facilitate the delivery of the permeant, alters the permeability of the skin, then this can significantly influence permeability. For example, certain solvents will alter the permeability of the skin barrier. These include water (whose hydration effects are, comparatively, somewhat limited), ethanol, propylene glycol and other organic solvents. These materials may be used to facilitate the delivery of drugs into or across the skin. For example, urea has also been shown to increase the hydration of the *stratum corneum* and increase the onset of erythema (Hellgren and Larsson 1979; Beastall et al. 1988). Urea has also been employed clinically to enhance hydrocortisone penetration from commercially available products, such as Alphaderm® and Calmurid®.

However, in the context of mathematical modelling, most of the data used to construct models come from experiments where formulation, other than solvents, is not considered (this is a limitation that will be discussed in detail later). Generally, the permeability data from which models have been constructed use aqueous or water–ethanol solvents in the donor phase, and few studies have explored formulation matters in detail (Pugh et al. 2005; Ghaforuain et al. 2010a, b; Moss et al. 2011). This also applies to ionisation, which is not generally considered by most models of percutaneous absorption.

The main issue with solvent effects in percutaneous absorption experiments that are relevant to the construction of mathematical models is the use of non-aqueous materials in the receptor compartment of the diffusion apparatus. Ideally, the receptor fluid should mimic as closely as possible the nature of the sink into which the permeant will diffuse when applied topically in vivo—the viable tissues of the skin and the blood supply. Chapter 1 described the structure of the skin and how its substantial vascularised structure provides a very efficient means of rapidly removing permeants and maintaining an effective diffusion gradient across the skin. Thus, an efficient receptor fluid in an in vitro experiment should similarly aim to maintain sink conditions and while not altering the nature of the membrane. In general, the vast majority of experiments use similar receptor fluids, although significant differences do occur and should be considered, both for their impact on permeation and for their impact on the subsequent, and in most cases unconnected, development of mathematical models. Most receptor fluids are based on buffered aqueous systems (or, in the case of non-ionised species, simple aqueous systems). Such systems are usually buffered at pH 7.4, usually with a phosphate buffer, to mimic physiological pH. However, there are exceptions, such as the use of a receptor phase buffered at pH 5 which is used to measure the permeability of aluminium across human skin (Mistry et al. 2013). This study used a receptor phase buffered at pH 5 as aluminium otherwise forms an insoluble oxide, which is difficult to analyse. Antimicrobial agents are often added to receptor fluids to maintain the integrity of the receptor compartment. If the penetrant of interest is lipophilic, then organic solvents, such as ethanol or propylene glycol, can be used, often in concentrations up to 25 %, with the remainder of the fluid being composed of water or a buffered aqueous system. Other solubilising agents have been used in receptor phases, including various surfactants and protein (i.e. bovine serum albumin). Care should be taken to consider the effect that the receptor compartment, through back-diffusion into the skin, may exert on barrier integrity, hence the suggestion to measure barrier integrity at the beginning and the end of an experiment. Therefore, the selection of a receptor fluid may be experiment-specific as different experiments aim to evaluate the permeability of a range of chemicals. Thus, care should be taken when using such information to construct models.

Detection of the Permeant

Once the receptor compartment fluid has been collected, it will be analysed by a number of commonly used methods. These include "cold" chromatographic methods (predominately high-performance liquid chromatography, HPLC, which is increasingly being coupled with mass spectrometric (MS) methods of analysis) or "hot" radiolabelled methods, usually by ^{14}C or ^{3}H-labelling of the permeant. This latter method is expensive, requires the use of licensed premises and, in the case of ^{3}H-labelled materials, offers the opportunity for tritium exchange between the material of interest and other materials in the analytical sample. HPLC conversely is comparatively inexpensive and widely used as a method of detection in biologically derived samples. In simpler diffusion experiments, such as those where artificial membranes (i.e. PDMS) have been used, UV spectrometry has been employed. While each method has its advantages and disadvantages, they all, if used and validated appropriately, can quantify the permeability of a material of interest such that accurate diffusion profiles can be produced, from which flux and permeability coefficients can be derived. Flux (as concentration/surface area/time, i.e. mg/cm^2/h) is usually the gradient of the steady-state portion of the diffusion profile. The permeability coefficient, k_p, is, in essence, the concentration-corrected flux and has units of distance and time (i.e. cm/s). Further, the lag time can be determined by extrapolating the steady-state section of the graph to the x-axis (the axis of the diffusion plot on which time is normally plotted). It is this data, mostly k_p but, in some cases flux (i.e. Magnusson et al. 2004), that is used by those constructing mathematical models of percutaneous absorption.

Conclusions

There exists a wide range of potentially conflicting experimental protocols for the determination of percutaneous absorption. All these methods are entirely valid in that they have been developed with specific penetrants in mind and not with the development of mathematical models, the latter activity often being unconnected to the original experiments. Thus, their results should be considered in this context—so too should their use in the development of the mathematical models and their limitations clearly enunciated and understood. For example, as we will see in later chapters, data sets have been compiled, and mathematical models developed, from experimental data that have been collated from different sources, which often use differing experimental protocols.

Ideally, in order to assess the relationships between the different methods, the relative permeabilities through the membranes of the same compounds would be assessed. This is, however, impractical as few comparative data exist for this purpose.

One solution to this problem has been to develop quantitative structure–activity (or permeability) relationships (QSARs, or QSPRs) based on the data that can be compared. QSPRs attempt to relate statistically the experimentally determined percutaneous penetration of a range of exogenous chemicals to known physicochemical parameters. However, such methods have been criticised as they, and their conclusions, are limited by the nature of the data on which they are based. This may include, for example, the lack of such models to consider formulation or ionisation effects. This may limit the scope of such models significantly. However, considered within this context, the models still yield valuable information on the mechanism of percutaneous absorption and, in many cases, allow the permeability of a chemical that is not within the data set but which fits within its "molecular space" to be predicted, provided that the model results in the output of a functional mathematical relationship between the experimentally measured parameter (usually k_p or J_{max}) and the statistically significant physicochemical properties of a penetrant.

Thus, this chapter has highlighted the complexity of a seemingly simple experiment from which permeability data are derived. It follows that the nature of this data, and its experimental origins in particular, should be described and understood when developing models, as the nature of the experimental data that forms the input to the mathematical model can clearly influence its output.

References

Anissimov YG, Roberts MS (1999) Diffusion modelling of percutaneous absorption kinetics. 1. Effects of flow rate, receptor sampling rate and viable epidermal resistance for a constant donor concentration. J Pharm Sci 88:1201–1209
Barry BW (1983) Dermatological formulations: percutaneous absorption. Marcel Dekker, New York
Bartek MJ, LaBudde JA, Maibach HI (1972) Skin permeability in vivo: comparison in rat, rabbit, pig and man. J Invest Dermatol 58:114–123
Beastall JC, Hadgraft J, Washington C (1988) Mechanism of action of Azone® as a percutaneous penetration enhancer: lipid bilayer fluidity and transition temperature effects. Int J Pharm 43:207–213
Berger A, Tanzella U, Machens HG, Liebau J (2000) Use of Integra in burn wounds and unstable scars. Chirurgia 71:558–563
Bronaugh RL, Maibach HI (1985a) Percutaneous Penetration. Marcel Dekker, London
Bronaugh RL, Maibach HI (1985b) Percutaneous-absorption of nitroaromatic compounds—in vivo and in vitro studies in the human and monkey. J Invest Dermatol 84:180–183
Bronaugh RL, Maibach HI (1999) Percutaneous absorption, 3rd edn. Marcel Dekker, Inc. CRC Press, New York
Bronaugh RL, Stewart RF, Congdon ER, Giles AL (1982) Methods for in vitro percutaneous absorption studies I: Comparison with in vivo results. Appl Pharmacol 62:481–488
Bronaugh RL, Hood HL, Kraeling MEK, Yourick JJ (1999) Determination of percutaneous absorption by in vitro techniques. In: Bronaugh RL, Maibach HI (eds.) Percutaneous absorption 3rd (Edn), New York: Marcel Dekker, Inc. CRC Press, pp 229–234
Bronaugh RL, Kraeling MEK, Yourick JJ, Hood HL (1999) Cutaneous metabolism during in vitro percutaneous absorption. In: Bronaugh RL, Maibach HI (eds.) Percutaneous absorption 3rd (Edn), New York: Marcel Dekker, Inc. CRC Press, pp 57–64
Bronaugh RL, Stewart RF, Simon M (1986) Methods for in vitro percutaneous absorption studies. 7. Use of excised human skin. J Pharm Sci 75:1094–1097
Chilcott RP, Barai N, Beezer AE, Brain SL, Brown MB, Bunge AL, Burgess SE, Cross S, Dalton CH, Dias M, Farinha A, Finnin BC, Gallagher SJ, Green DM, Gunt H, Gwyther RL, Heard CM, Jarvis CA, Kamiyama F, Kasting GB, Ley EE, Lim ST, McNaughton GS, Morris A, Nazemi MH, Pellett MA, Du Plessis J, Quan YS, Raghavan SL, Roberts M, Romonchuk W, Roper CS, Schenk D, Simonsen L, Simpson A, Traversa BD, Trottet L, Watkinson A, Wilkinson SC, Williams FM, Yamamoto A, Hadgraft J (2005) Inter- and intra-laboratory variation of in vitro diffusion cell measurements: an international multicenter study using quasi-standardised methods and materials. J Pharm Sci 94:632–638
Chow C, Chow AYK, Downie RH, Buttar HS (1978) Percutaneous absorption of hexachlorophene in rats, guinea pigs and pigs. Toxicology 9:147–154
Dias MMR, Raghavan SL, Pellett MA, Hadgraft J (2003) The effect of beta-cyclodextrins on the permeation of diclofenac from supersaturated solutions. Int J Pharm 263:173–181
Elias PM, Fritsch P, Tappeine G, Mitterma H, Wolff K (1974) Experimental staphylococcal toxic epidermal necrolysis (10) in adult humans and mice. J Lab Clin Med 84:414–424
Esposito E, Zanella C, Cortesi R, Menegatti E, Nastruzzi C (1998) Influence of liposomal formulation parameters on the in vitro absorption of methyl nicotinate. Int J Pharm 172:255–260
Franz TJ (1975) On the relevance of in vitro data. J Invest Dermatol 64:190–195
Friend DR (1992) In vitro skin permeation techniques. J Cont Rel. 18:235–248
Geinoz S, Rey S, Boss G, Bunge AL, Guy RH, Carrupt PA, Reist M, Testa B (2002) Quantitative structure-permeation relationships for solute transport across silicone membranes. Pharm Res 19:1622–1629
Ghafourian T, Samaras EG, Brooks JD, Riviere JE (2010a) Modelling the effect of mixture components on permeation through skin. Int J Pharm 398:28–32

Ghafourian T, Samaras EG, Brooks JD, Riviere JE (2010b) Validated models for predicting skin penetration from different vehicles. Eur J Pharm Sci 41:612–616

Hager DF, Mancuso FA, Nazareno JP, Sharkey JW, Siverly JR (1994) Evaluation of a cultured skin equivalent as a model membrane for iontophoretic transport. J Cont Rel 30:117–123

Harada K, Murakami T, Kawasaki E, Hagashi Y, Yamamoto S, Yata N (1993) In vitro skin permeability to salicyclic acid of human, rodent and shed snake skin. J Pharm Pharmacol 45:414–418

Harrison SM, Barry BW, Dugard PH (1984) Effects of freezing on human skin permeability. J Pharm Pharmacol 36:261–262

Hellgren L, Larsson K (1979) On the effect of urea on human epidermis. Dermatologica 149:289–293

Horiguchi Y, Suzuki S, Toda K, Ueda M, Leigh IM, Imamura S (1997) Three-dimensional cultures of keratinocytes and an application to in vitro amyloid production of cutaneous amyloidosis. Microsc Res Tech 38:387–393

Imhof RE, de Jesus MEP, Xiao P, Ciortea LI, Berg EP (2009) Closed-chamber transepidermal water loss measurement: microclimate, calibration and performance. Int J Cosmet Sci 31:97–118

Itoh T, Xia J, Magavi R, Nishihata T, Rytting JH (1990) Use of shed snake skin as a membrane for in vitro percutaneous penetration studies. Pharm Res 7:1042–1047

Kligman AM, Christophers E (1963) Preparation of isolated sheets of human stratum corneum. Pharmacol Skin 1:138–153

Kremer M, Lang E, Berger AC (2000) Evaluation of dermal-epidermal skin equivalents ("composite-skin") of human keratinocytes in a collagen-glycosaminoglycan matrix (Integra™ Artificial Skin). Br J Plast Surg 53:459–465

Kuroski Y, Nagahara N, Taniza WA, Nishmura H, Nakayama T, Kimura T (1991) Use of lipid disperse systems in transdermal drug delivery—comparative study of flufenamic acid permeation among rat abdominal skin, silicon rubber membrane and stratum corneum sheet isolated from hamster cheek pouch. Int J Pharm 67:1–9

Lam LT, Sun Y, Davey N, Adams R, Prapopoulou M, Brown MB, Moss GP (2010) The application of feature selection to the development of Gaussian Process models for percutaneous absorption. J Pharm Pharmacol 62:738–749

Lin SY, Hou SJ, Hsu THS, Yeh FL (1992) Comparisons of different animal skins with human skin in drug percutaneous absorption studies. Meth Find Exp Clin Pharmacol 14:645–654

Machens HG, Berger AC, Mailaender P (2000) Bioartificial skin. Cells Tissue Organs 167:88–94

Magnusson BM, Anissimov YG, Cross SE, Roberts MS (2004) Molecular size as the main determinant of solute maximum flux across the skin. J Invest Dermatol 122:993–999

Maibach HI (1975) Animal models in dermatology. Churchill-Livingstone, Edinburgh

Mali JHW (1978) Current problems in dermatology, vol 7. Karger, Basel

Martin A, Swarbrick J, Cammarata A (1983) Physical pharmacy, 3rd edn. Lea & Febinger, Philadelphia

Marzulli FN, Maibach HI (1975) Relevance of animal models: the hexachlorophene story. In: Maibach HI (ed) Animal models in dermatology. Churchill-Livingstone, Edinburgh

Megrab NA, Williams AC, Barry BW (1995a) Oestradiol permeation across human skin, silastic and snake skin membranes: the effects of ethanol/water co-solvent systems. Int J Pharm 116:101–112

Megrab NA, Williams AC, Barry BW (1995b) Oestradiol permeation through human skin and silastic membrane: effects of propylene glycol and supersaturation. J Cont Rel 36:277–294

Minghetti P, Casiraghi A, Cilurzo F, Montanari L, Marazzi M, Falcone L, Donati V (1999) Comparison of different membranes with cultures of keratinocytes from man for percutaneous absorption of nitroglycerine. J Pharm Pharmacol 51:673–678

Mistry T, Ajum K, Owen J, Stair J, Wilkinson SC, Staff K, Moss GP (2013) The percutaneous absorption of aluminium from antiperspirant products. In: Brain KR, Chilcott R (eds) Advances in the dermatological sciences. UK, Royal Society of Chemistry, Cambridge, pp 256–264

Mizunuma M, Yanai A, Seno H, Hirabayashi S (2000) Experience in repair utilizing artificial skin for exposed bone surfaces. Eur J Plast Surg 23:305–308

Moss GP, Gullick DR, Cox PA, Alexander C, Ingram MJ, Smart JD, Pugh WJ (2006) Design, synthesis and characterisation of captopril prodrugs for enhanced percutaneous absorption. J Pharm Pharmacol 58:167–177

Moss GP, Sun Y, Davey N, Adams R, Pugh WJ, Brown MB (2009) The application of Gaussian Processes to the prediction of percutaneous absorption. J Pharm Pharmacol 61:1147–1153

Moss GP, Sun Y, Wilkinson SC, Davey N, Adams R, Martin GP, Prapopoulou M, Brown MB (2011) The application and limitations of mathematical modelling in the prediction of permeability across mammalian skin and polydimethylsiloxane membranes. J Pharm Pharmacol 63:1411–1427

Nemecek GM, Dayan AD (1999) Safety evaluation of human living skin equivalents. Tox Pathol 27:101–103

Pelle E, Mammone T, Combatti M, Marenus K, Maes D (1993) UV-induced lipid-peroxidation in living skin equivalents. Clin Res 41:A504

Potts RO, Guy RH (1992) Predicting skin permeability. Pharm Res 9:663–669

Pugh WJ, Wong R, Falson F, Michniak BB, Moss GP (2005) Discriminant analysis as a tool to identify compounds with potential as transdermal enhancers. J Pharm Pharmacol 57:1389–1396

Raghavan SL, Kiepfer B, Davis AF, Kazarian SG, Hadgraft J (2001) Membrane transport of hydrocortisone acetate from supersaturated solutions; the role of polymers. Int J Pharm 221:95–105

Raghavan SL, Schuessel K, Davis A, Hadgraft J (2003) Formation and stabilisation of triclosan colloidal suspensions using supersaturated systems. Int J Pharm 261:153–158

Ramsamooj R, Strande L, Kain MH, Doolin EJ, Hewitt CW (1998) A novel in vitro model for xenorejection and immune mechanisms using bioengineered living skin equivalents. Transplant Proc 30:1087–1088

Roberts MS, Cross SE (1999) A physiological pharmacokinetic model for solute disposition in tissues below a topical application site. Pharm Res 9:1392–1398

Roberts ME, Mueller KR (1990) Comparisons of in vitro nitroglycerin (TNG) flux across Yucatan pig, hairless mouse and human skins. Pharm Res 7:673–676

Roberts MS, Anissimov YG, Gonsalvez RA (1999) Mathematical models in percutaneous absorption. In: Bronaugh RL, Maibach HI (eds.) Percutaneous absorption 3rd (Edn), New York: Marcel Dekker, Inc. CRC Press, pp 3–56

Sato K, Sugibayashi K, Morimoto Y (1991) Species differences in percutaneous absorption of nicorandil. J Pharm Sci 80:104–107

Stott PW, Williams AC, Barry BW (2001) Mechanistic study into the enhanced transdermal permeation of a model β-blocker, propranolol, by fatty acids: a melting point depression effect. Int J Pharm 219:161–176

van Hal D, van Rensen A, de Vringer T, Junginger H, Bouwstra J (1996) Diffusion of estradiol from non-ionic surfactant vesicles through human stratum corneum in vitro. STP Pharma Sci 6:72–78

Wang CK, Nelson CF, Brinkman AM, Miller AC, Hoffler WK (2000) Spontaneous cell sorting of fibroblasts and keratinocytes creates an organotypic human skin equivalent. J Invest Deramtol 114:674–680

Wester RC, Maibach HI (1976) Relationship of total dose and percutaneous absorption in Rhesus monkey and man. J Invest Dermatol. 67:518–520

Wester RC, Noonan PK (1980) Relevance of animal models for percutaneous absorption. Int J Pharm 7:99–110

Williams AC (2003) Transdermal and topical drug delivery. The Pharmaceutical Press, London

Woolfson AD, McCafferty DF, Moss GP (1998) Development and characterisation of a moisture-activated bioadhesive drug delivery system for percutaneous local anaesthesia. Int J Pharm 169:83–94

Yang EK, Yoon HH, Lee DH, Park JK (2000) Assessment of toxic potential of industrial chemicals using a cultured human bioartificial skin model: production of interleukin 1α and hydroxyeicosatetraenoic acids. Skin Pharmacol App Skin Physio 13:246–257

Young CD, Wu JR, Tsou TL (1998) Fabrication and characteristics of polyHEMA artificial skin with improved tensile properties. J Mem Sci. 146:83–93

Zhai HB, Maibach HI (2001) Effects of skin occlusion on percutaneous absorption: an overview. Skin Physiol Appl Skin Pharmacol 14:1–10

Chapter 3
Mathematical Treatments and Early Models of Skin Permeability

Introduction

Most scientific researchers will attempt to quantify their findings in mathematical terms, or to derive methods that allow predictive estimates to be made with their data. Such techniques are common to a range of disciplines, and in the context of percutaneous absorption, this began with mathematical models of skin absorption and has progressed to statistical models that mechanistically describe this process for a given data set and which also, in most cases, allow the prediction of percutaneous absorption for chemicals not included in the data set. While the aim of this book is to focus on the latter it must do so by describing the former, and this chapter describes how various mathematical models of percutaneous absorption underpinned our burgeoning understanding of skin permeation and the development of early quantitative models of percutaneous absorption.

Despite the complex, heterogenous nature of the skin—and its inherent biological variation—it has been mathematically represented in a general sense by Fickian diffusion, and in particular Fick's second law of diffusion. This comparatively simple model describes the vast majority of skin permeation and, as such, will be the focus of this chapter. In addition, as the vast majority of models for percutaneous absorption are based on data from in vitro infinite-dose experiments, this chapter will focus mainly on that aspect of the subject. Where this is not the case these experiments, and the mathematics associated with them, are discussed in Chap. 8. Further, for a more comprehensive discussion of this subject, the reader is directed towards the excellent chapter by Roberts et al. (1999).

It is well over one hundred years since it was first proposed that biological activity could be influenced by a physical property—this being, in the case of Crum-Brown and Fraser's work, the connection between aqueous solubility and physiological action (Crum-Brown and Fraser 1868). Similarly, Richardson (1869) demonstrated the link between *n*-alcohols and their physiological action, and, some years later, Richet (1893) demonstrated the link between lipophilicity (or the inverse link

G.P. Moss et al., *Predictive Methods in Percutaneous Absorption*,
DOI 10.1007/978-3-662-47371-9_3

between aqueous solubility) and toxicity of a range of simple polar chemicals, including ethers, alcohols and ketones. The theories of Overton (1897) and Meyer (1899) elaborated on the above findings, proposing that the partition of a chemical between water and oil phases simulated the relationship between a chemical and its interaction with a lipophilic receptor and its surrounding aqueous environment. Berthelot and Jungfleisch (1872) defined the partition coefficient as the ratio of concentrations—at equilibrium—of a solute between immiscible aqueous and lipid phases. Fundamentally, therefore, it could be said that the work of Meyer and Overton in particular implied that, the more lipophilic a compound, the better it would penetrate into a lipid membrane. This has been elaborated on in Chap. 1, in the context of skin permeability. It is also interesting to contextualise the above comments with Ferguson's (1939) postulation on toxic doses being a fixed fraction of a chemical's aqueous solubility, or is at a constant thermodynamic activity. Extrapolating this principle from toxicity to a biological response—in the current context, skin permeability—suggests that the degree of response will change as solubility of penetrants is varied, or has practical limitations depending on a particular context (such as the solubility profile across the *stratum corneum*). Clearly, such changes are not linear in nature and, in consideration of Meyer and Overton's theory in particular, simply increasing the solubility of a potential penetrant ad infinitum will not always result in desired improvements in solubility—or, as Dearden (1985) put it, in describing changes in chemical structure: "methyl, ethyl, butyl, futile".

Significant advances in predicting percutaneous absorption, both theoretically and mathematically, have been made since the 1940s. For example, Rothman (1943) illustrated the importance of both penetrant solubility and the interaction of the penetrant with its vehicle. Hadgraft and Somers (1956) determined that a balance of lipophilic and hydrophilic properties (ideally, a log[water:lipid] partition coefficient of between 1 and 3) was necessary for successful percutaneous permeation, reflecting the comparative lipophilic and hydrophilic natures of the *stratum corneum* and the underlying viable epidermis and dermis.

Higuchi (1960, 1961) derived a mathematical model of percutaneous permeation based on Fick's first law. This model describes the process of percutaneous absorption as a passive diffusion process in a serial model, which includes the vehicle/formulation and the different layers of the skin. Higuchi also emphasised the importance of thermodynamic activity of the penetrant to the permeation process which was applied to modelling the permeation of the active ingredient from topical ointments and which resulted in the publication of key equations that described the fundamental relationships between release rates, drug concentrations and diffusion coefficients.

Therefore, this period saw the development of key principles in the percutaneous absorption of exogenous chemicals—that both partitioning and solubility were significant factors in modelling or determining skin permeation. It also led onto an increasing range of models that described in mathematical terms the process of skin permeation of an exogenous chemical. One such example is the development of a range of pharmacokinetic models of skin permeation. While Roberts et al. (1999) reviewed these models in considerable detail, it is worth discussing one of them—

by Guy and Hadgraft (1983)—in more detail as it built on the key developments in our understanding of skin permeability in the context of a compartmental model that reflects the structure of the skin—or, more broadly, the structure of the "path to the receptor" via the skin (Aulton 2003). The aim of any mathematical model for percutaneous absorption is to provide an estimate of drug levels in each "compartment"—which may include the vehicle but is more concerned with what may partition into the various skin layers, the subcutaneous fatty tissue and blood after application of a transdermal drug delivery device. Thus, various attempts have been made at using pharmacokinetic modelling as an alternative to diffusion-based models (Riegelman 1974, shown in Fig. 3.1; Chandraskeanan 1978). Wallace and Barnett (1978) used a series of multicompartment models to describe the percutaneous absorption of methotrexate, which yields the expression:

$$A_3 = \frac{A_1 k_{12} k_{23}}{k_{21} + k_{23}} \left(t - \frac{1}{k_{21} + k_{23}} \right) \tag{3.1}$$

where the terms used are shown in Fig. 3.1. This expression indicates that, at long times, there is a steady-state flux of $(A_1 k_{12} k_{23})/(k_{21} + k_{23})$ and a lag time of $1/(k_{21} + k_{23})$, where $A_1 k_{12} = k_p C_v$ (Wallace and Barnett 1978; Roberts et al. 1999).

Guy and Hadgraft (1982) used a pharmacokinetic model to describe in vivo blood and urine levels after topical application, and to relate this to the physical processes governing absorption. This is shown in Fig. 3.2 and illustrates the four-component pharmacokinetic model derived by Guy and Hadgraft (1982). It describes the process of drug diffusion through the *stratum corneum*, k_1, and the epidermis, k_2. Additionally, this model takes into account the reservoir effect of the *stratum corneum* by measuring the relative affinities of the penetrant for the *stratum corneum* and viable epidermis, represented by k_3. Finally, k_4 portrays the classical pharmacokinetic rate constant following intravenous administration. This model may be further modified by taking into account the vehicle effects, and as such may be used to screen potential transdermal formulation candidates. For example, it was applied to a study of the topical administration of sunscreens by Walters et al. (1998), and Guy et al. (1983) used it to estimate plasma levels resulting from multiple dosing of topically applied formulations. This model was modified by Guy and Hadgraft (1984) to include epidermal metabolism (notated by k_5 and k_6, which represented a metabolic step and the removal of the metabolite, respectively) by the inclusion of an additional compartment in the model.

In general, mathematical models relating to the specific penetration and absorption of a particular chemical may require a specific model due to the physicochemical properties of the penetrant and its specific pharmacokinetic and pharmacodynamic profile in the body. One example is the penetration of a local anaesthetic drug, which may be somewhat simplified as the site of action is at or near the dermo-epidermal junction. Thus, the resistance from the dermis may therefore be neglected, and systemic uptake is not relevant to the efficacy of such

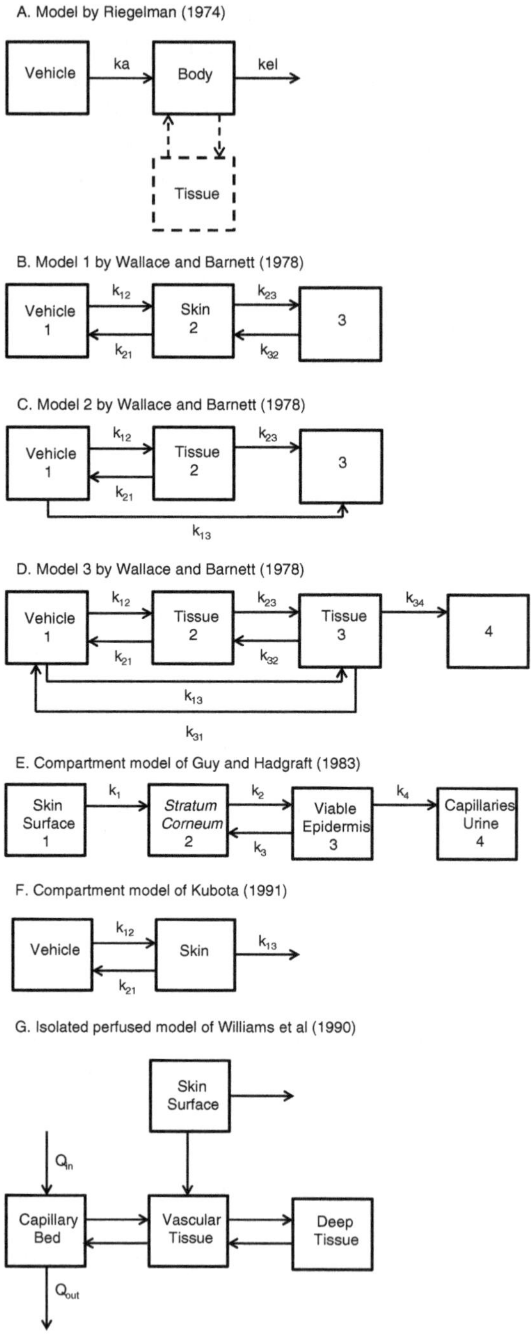

Fig. 3.1 Examples of compartmental models of skin permeation (modified from Roberts et al. 1999)

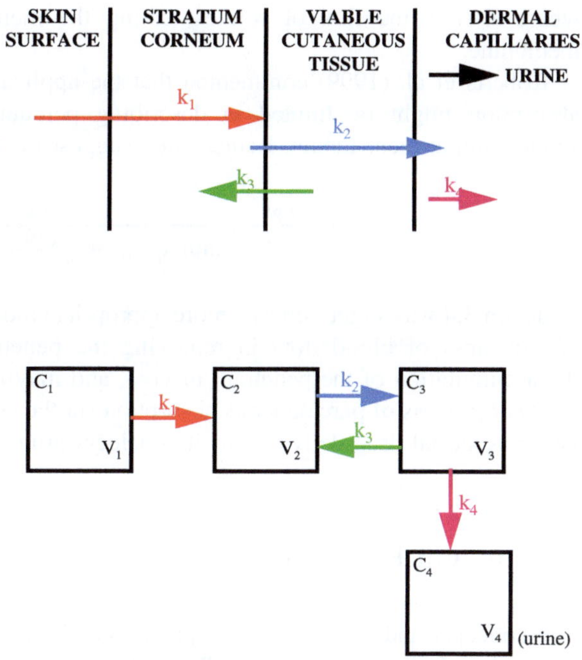

Fig. 3.2 *Four-compartment pharmacokinetic model* (modified from Guy and Hadgraft 1982)

drugs. The total resistance of the skin, represented by R, to percutaneous absorption of a local anaesthetic can be described by:

$$R = \frac{h}{F_{SC} \cdot D_{SC} \cdot K_{SC}} + \frac{h}{F_e \cdot D_e \cdot K_e} \tag{3.2}$$

where

h the thickness of the resistor membrane;
F fractional area of the route, where there is more than one pathway involved;
D diffusion coefficient of the drug;
K partition coefficient of the drug;

and the denominator subscripts sc and e refer to the *stratum corneum* and the epidermis, respectively (Woolfson and McCafferty 1993).

It should also be noted that diffusion-based pharmacokinetic models, such as that described by Cooper (1976), have been applied to percutaneous absorption. Cooper assumed diffusion through the skin according to the flux of solutes exiting the membrane per unit area, $J_s(t)$, and used a single-compartment model to do so:

$$J_s(t) = -D_m \frac{\partial C_m}{\partial x} |x = h_m = k_p C_v \left[1 + 2 \sum_{n=1}^{\infty} (-1)^n . \exp\left(-\frac{t}{t_d} \pi^2 n^2 \right) \right] \tag{3.3}$$

where $J_s(t)$ is the flux of solutes exiting the membrane per unit area of the membrane.

Roberts et al. (1999) commented that the application of this model to in vivo absorption might be limited in describing percutaneous absorption as it is a single-compartment model. Rather, they suggest the following expression:

$$\widehat{J}s(s) = \frac{k_p C_v}{s} \cdot \frac{\sqrt{st_d}}{\sinh(\sqrt{st_d}) + \frac{\sqrt{st_d}}{st_d\lambda_2 + \lambda_1} . \cosh(\sqrt{st_d})} \tag{3.4}$$

Equation 3.4 was suggested as a more appropriate model as it takes into account the effectiveness of blood flow in removing the penetrant from the epidermis and the accumulation of the penetrant in vivo, and may be more appropriate to model the *total* process of percutaneous absorption (in the context of systemic absorption, i.e. transdermal drug delivery), as it considers drug absorption and elimination.

Infinite and Finite Dosing

The concepts and experimental requirements for infinite and finite doses are discussed briefly in Chap. 2 and will also be discussed in greater detail in Chap. 8, where examples of finite-dose models will be discussed. They are also outlined below in the context of their underpinning mathematics.

An infinite-dose experiment is one where the concentration of permeant in the donor compartment remains constant and dose not therefore lessen over the duration of the experiment. Examples of infinite-dose experiments include those where saturated or supersaturated solutions, or transdermal patches, are applied to the skin surface. On the other hand, finite-dose experiments are those where a small amount of the penetrant of interest is applied to the skin surface and which will diminish during the course of the experiment. Examples of finite-dose experiments include those that consider the application of pharmaceutical creams and ointments or the application under non-occluded conditions formulations that include volatile components (normally solvents).

Transport of exogenous chemicals into and across the skin is characterised by its flux (J), that is, the concentration per unit area per unit time that passes across the skin. To date, no active transport mechanism has been identified for this process and it is understood to progress by passive diffusion due to a thermodynamic force based on a concentration gradient of the penetrant either side of the membrane across which diffusion occurs.

In the context of the predictive modelling of skin absorption, quantitative structure-activity (or permeability) relationships (QSARs or QSPRs) aim to relate statistically the biological activity of a range of penetrants (the data set from which the model is constructed) to the structural and physicochemical properties of penetrants and their experiments (Sinko 2010). More recently, as is described in

Chap. 9, experimental parameters may additionally be added to models. Such processes result in models that provide two main functions: as they are normally in the form of an algorithm (i.e. an equation), they can offer mechanistic insight to the key structural and physicochemical parameters that influence the absorption process, and they can allow, by substitution of the data for a new chemical into the derived equation, a prediction of the permeability for that chemical. However, it may be the case that an algorithm featuring specific, discrete parameters suggests that such parameters do not exhibit any degree of overlap or covariance—as, for example, we see with molecular weight and melting point as we increase the carbon number of a series of *n*-alcohols. Therefore, the potentially composite or covariate nature of some descriptors employed in modelling studies should be considered when interpreting the output of models.

The aim of this book, and more specifically this chapter, is to consider the experimental and mathematical context in which quantitative structure–permeability relationships are derived. This chapter will focus on the mathematics underlying in vitro infinite-dose experiments, as the vast majority of these models are based on data sets constructed from experiments of that nature, where conditions of pseudo-steady-state permeation exist. While the system we are interested in is biological in nature, we are not examining a biological process, such as protein binding, binding to a receptor or active transport, but rather a passive diffusion process in a biological system. Nevertheless, as we have seen with some of the examples above, the overall diffusion process should be considered in the context of pharmacokinetics and metabolism, something which simple models/algorithms of permeability do not currently incorporate.

The models described above are extremely complex and relate to a range of circumstances, including finite and infinite doses and also consider the adsorption–distribution–metabolism–elimination (ADME) process in the context of percutaneous absorption. They have been described considerably in the excellent article by Roberts et al. (1999). However, in the context of algorithms for percutaneous absorption, a significant simplification of such models is often considered and relates to the application of principles of Fickian diffusion to the problem domain of percutaneous absorption.

This, Fick's first law has been used to describe the transfer of a diffusing material through a membrane (Barry 1983):

$$J = -D\frac{\partial C}{\partial x} \tag{3.5}$$

where

J is the flux of the permeant (the rate of transfer per unit area per unit time);

C is the concentration of the diffusing substance;

x is the spatial coordinate measured normal to the section (meaning that $\partial C/\partial x$ is the concentration gradient across the membrane); and

D is the diffusion coefficient, or the diffusivity of the penetrant.

According to Blank (1969), this model may only be applied to relatively low concentrations of penetrant, as deviations from Fick's first law become apparent when the concentration increases greatly.

From Eq. 3.5, Fick's second law of diffusion can be derived. It is an equation of mass transport, emphasising the change in concentration, not mass, per unit time diffusing across a barrier membrane. It assumes that penetration is unidirectional which, in an isotropic medium, may be given by:

$$\frac{\partial C}{\partial t} = D \cdot \frac{\partial^2 C}{\partial x^2} \tag{3.6}$$

where t represents time.

Equation 3.6 may also be written in a more general form to represent three dimensions:

$$\frac{\partial C}{\partial t} = D \cdot \left(\frac{\partial^2 C}{\partial x^2} + \frac{\partial^2 C}{\partial y^2} + \frac{\partial^2 C}{\partial z^2} \right) \tag{3.7}$$

However, most diffusional movement is in one direction only, and Eq. 3.5 adequately describes most experimental diffusional cases (Tojo 1987).

Fick's second law states that "the change in concentration with time in a particular region is proportional to the change in the concentration gradient at that point in the system" (Martin et al. 1983). Solving Fick's second law is, however, difficult and depends on boundary conditions for a particular experiment. For example, Franz (1975), using his eponymous diffusion cells and an infinite-dose experimental protocol, modified Fick's equation and represented the amount of drug in the donor phase with the following expression:

$$\log \frac{M}{t^{3/2}} = \log \left(\frac{8C}{h^2 \pi^{1/2}} \right) + \left(\frac{3}{2} \log D \right) - \left(\frac{h^2}{9.2 \mathrm{D} t} \right) \tag{3.8}$$

where

M cumulative amount of drug in receptor phase;
D diffusion coefficient of drug in skin;
h thickness of skin;
C concentration of drug in donor compartment; and
t time elapsed in the experiment.

However, while Eq. 3.8 has been successfully applied by other researchers (for example, Albery and Hadgraft 1979), simpler expressions are generally more frequently employed in both experimental and modelling experiments of percutaneous absorption.

Thus, following from the comments above, descriptions in Chap. 1 of the complex nature of the skin and, in Chap. 2, of the range of experiments (including the use of different mammalian skin tissues, or parts of skin tissue such as isolated *stratum corneum* or dermatomed skin, or even artificial membranes such as polydimethylsiloxane) used for particular experiments, it follows that the application of Fick's law to diffusion offers, at best, an approximation of the diffusion process across the skin, a comment which should be considered in the wider context of how closely in vitro experiments can mimic in vivo conditions and—in the wider context of this book—how accurate models of absorption can be if they, as their theoretical basis, use models based on simplifications and approximations. As was highlighted in Chap. 2, the manner in which this is achieved usually relates to an experiment where the membrane of interest (i.e. whole skin, isolated *stratum corneum*, epidermis, dermatomed skin or an artificial membrane) sits between two compartments: the donor compartment from where the chemical of interest will diffuse and the receptor compartment into which it will diffuse. After a period of equilibration (often referred to as the "lag phase" of the diffusion process) steady-state diffusion will be achieved when the concentration gradient across the membrane is constant, in which case Fick's second law (Eq. 3.6) may be written as:

$$\frac{\partial M}{\partial t} = \frac{DC_0}{h} \tag{3.9}$$

where

M is the cumulative mass of permeant passing across the membrane per unit area per unit time (t);

C_0 is the concentration of the permeant at the skin surface; and

h is the thickness of the membrane.

C_0 is, in practice, difficult to measure, and the concentration in the donor solution or formulation, C_V, is normally used as a readily measurable approximation. C_0 and C_V are related by P, the partition coefficient of the permeant between the vehicle and the membrane:

$$P = \frac{C_0}{C_V} \tag{3.10}$$

and

$$C_0 = PC_V \tag{3.11}$$

Substitution of Eq. 3.8 into Eq. 3.6 gives the following expression:

$$\frac{\partial M}{\partial t} = \frac{DPC_V}{h} \tag{3.12}$$

Thus, a plot of M (the cumulative amount of permeant, passing across a defined area of the membrane (as for example, μg/cm^2 against time)) results in a profile of permeation, from the donor compartment, across the membrane and into the receptor compartment. The lag time can be determined by extrapolating the pseudo-steady-state area of the diffusion profile to the x(time)-axis, or after the method derived by Crank (1975):

$$L = \frac{h^2}{6D} \tag{3.13}$$

where L is the lag time.

This equation infers that the diffusion coefficient can be determined directly from the lag time. However, while this holds for simple isotropic membranes, it does not apply to more complex membranes, such as mammalian skin. Further, Williams (2003) also commented that lag times obtained from experiments are highly variable and may be influenced by permeant-skin binding. An alternative approach is to rearrange Eq. 3.9:

$$D = \frac{{}^{\partial M}/_{\partial t} . h}{PC_{\mathrm{V}}} \tag{3.14}$$

Given that is the gradient of the diffusion profile—the flux (J)—this expression may be simplified further:

$$D = \frac{J.h}{PC_{\mathrm{V}}} \tag{3.15}$$

Equation 3.15 allows the calculation of the diffusion coefficient and, while still influenced by experimental variation, it is lessened as this expression does not contain squared terms, as Eq. 3.10 does, which comparatively reduces any error.

The main focus of QSPRs for skin permeability has not been the flux, but the permeability coefficient, k_{p}. The permeability coefficient is the steady-state flux normalised by the concentration gradient across the membrane. This may be expressed by:

$$k_{\mathrm{p}} = \frac{J_{\mathrm{ss}}}{\Delta C_{\mathrm{V}}} \tag{3.16}$$

where

J_{ss} is the steady-state flux determined from the diffusion profile (with units of concentration per unit area per unit time);

ΔC_{V} is the concentration gradient across the membrane.

The concentration gradient is maximised if the concentration below the membrane is essentially zero—in practice, this means maintaining sink conditions where possible. It also indicates, as discussed in Chap. 2 in the context of permeability experiments, how an alteration in the concentration gradient (by, for example, failing to maintain sink conditions or in the case of solvent loss in some finite-dose experiments) can influence the rate of diffusion and the flux across the skin. This is particularly important in the practical sense of maximising permeation across the skin of, for example, pharmaceutically active materials. Further, Crank (1975), in defining the skin as a single pseudo-homogenous membrane, suggested that Eqs. 3. 14 and 3.16 may be rearranged to provide the following expression:

$$k_p = \frac{PD}{h} \tag{3.17}$$

Substituting into Eq. 3.12 this gives the following expression:

$$\frac{\partial M}{\partial t} = J = k_p C_V \tag{3.18}$$

The flux may be obtained from the gradient of the diffusion profile, being the gradient of the steady-state component of the concentration–time plot of permeant diffusion. k_p may then be determined as the concentration gradient is known or can be approximated based on skin conditions. It should be noted that k_p and C_v should be determined in the same solvent (or vehicle) and that models cannot be developed using mixtures of solvents (i.e. with k_p determined for an aqueous solvent and C_v for a non-aqueous solvent). The permeability coefficient is often used to characterise skin permeability in preference to other parameters as is it relatively straightforward to determine and, as it normalises for concentration, allows comparisons of permeability for a range of permeants to be made. However, it is clear that J_{max} is a more practical and useful term to use, particularly from pharmaceutical or toxicological perspectives, being used, for example, as a key characteristic of the clinical performance of a transdermal drug delivery device. Thus, in terms of QSPR models, skin absorption, as k_p, is the most widely used term, although some significant studies, most notably Magnusson et al. (2004), derive models based on steady-state flux. As described in Chap. 2, it is a limitation of these models that they seldom account for formulation effects. However, k_p is used as, over time, a large body of data has been published in this format, which allows comparisons of theoretical models with physical (i.e. experimental or clinical) measurements, and Kroes et al. (2007) have considered the use of J_{max} and k_p in the context of risk assessment for cosmetic formulations.

It is a further disadvantage of this approach that some of the parameters, such as D/h and concentration terms (in, for example, the *stratum corneum*), are impractical if not downright impossible to measure experimentally with any degree of accuracy. Kasting et al. (1992) proposed that a penetrant's saturated solubility—in the vehicle in which it contacts the skin, $C_{v,\ sat}$—may be calculated from ideal solution theory or estimated from its saturated solubility in model organic solvents, such as octanol, that are used to mimic the *stratum corneum*. Further, they represented the *stratum corneum* in a simplified manner by assuming that its lipids were highly ordered, thus treating this barrier as a semisolid, akin to the structure and order associated with polymeric membranes, such as polydimethylsiloxane, that find use as skin substitutes in in vitro experiments (see Chap. 2). By doing so, they then considered that transport into and across this membrane was via a "free volume" mechanism, which allowed the diffusivity of the *stratum corneum* to be modelled. Potts and Guy (1992) then developed from this an expression for skin permeability:

$$\log k_p = \log\left(\frac{D_0}{h}\right) + \log K - \left[\frac{\beta . V}{2.303}\right] \tag{3.19}$$

where V is the molecular volume of the penetrant, and D_0 and β are constants. They presented this expression in a more general form as:

$$\log k_p = a + b.\log P - c.\mathrm{MW} \tag{3.20}$$

In doing so, they made several assumptions, including that the partition coefficient could be represented by a function of P, and that molecular volume could be approximated by molecular weight (Potts and Guy 1992). This provides a wide range of readily available information on which to build models, as values of P/log P are widely available for thousands of chemicals in the literature (or can be calculated by a range of software packages which use a range of methods to determine the values).

Attempts to develop a "global" or "holistic" approach to the use of QSAR-based models in predicting skin permeability was assisted greatly by the publication of 97 permeability coefficients (for 94 different chemicals) (Flynn 1990). These data related mostly to in vitro percutaneous absorption experiments using human skin, except for in vivo studies (for toluene, ethylbenzene and styrene). This provided the largest data set of collated human skin permeability data and was widely used—in its original form as presented by Flynn, by the addition of new chemicals subsequently published or by the abstraction of specific subsets of chemically related data—by a number of researchers over the next ten years or so (see Chap. 4). Flynn's data were collated from fifteen different literature sources, which infer that the data inevitably exhibit a high degree of experimental variation which may be due to interlaboratory variation, including the use of skin from different sources and locations on the body of those donating it. The publication of Flynn's data set was a substantial milestone in the development of QSARs for the prediction of skin permeability and forms the basis of the work reviewed in this volume.

Flynn's study was developed in the context of the skin permeability of pollutants and their attendant risk to human health. He considered the "permeation triangle", consisting of exogenous chemicals, or pollutants as one corner, the permeability of skin as the second and water as the third corner of his triangular model, both in terms of its role in the skin permeation process and its role as a vehicle for pollutants, or potential permeants. The aqueous solubility of permeants therefore sets the upper limit on absorption, and therefore, consideration of permeation from saturated aqueous solutions was a key reference point. Further, the interaction between water and the skin, particularly regarding hydration and changes to barrier function, was central to his model particularly during long exposures and diffusive mobility within the skin during the absorption process.

Another important concept addressed by Flynn is the limitation of in vitro laboratory diffusion models in producing a model for in vivo absorption processes. He very clearly identified the need for a working model to assess permeation (as risk, or pharmaceutical efficacy). Diffusional pathways through the skin had been identified, but Flynn commented that the transepidermal and transfollicular routes were difficult to factor. Thus, parallel routes of absorption cannot be differentiated based on specific pathways and any model produced does not need to specify separate diffusional pathways, instead being merged and averaged into a single "top-down" model of skin absorption. However, Flynn was clear that the available data did, on inspection, demonstrate the significance of polarity in the permeation process and that provision should be made for an aqueous pore pathway and a non-polar pathway in any resultant model.

Thus, in Flynn's model, diffusion across the horny layer, or via the transfollicular route, and into the local circulation is viewed as permeation of aqueous pores by polar substances or as diffusional passage through an isotropic aqueous layer, with both routes reflecting the average properties of the skin. A consistent message from Flynn's work is the reiteration that any such model is, by its very nature, an approximation of "real-world" skin permeation, and cannot consider in detail issues such as clearance and the size and nature of the diffusional sink.

Thus, Flynn's model has two functional strata in series: the total diffusional resistance of the *stratum corneum* and the viable tissue (or their follicular counterparts):

$$R_T = R_{sc} + R_{vt} \tag{3.21}$$

where
R_T is the total diffusional resistance;
R_{sc} is the diffusional resistance of the *stratum corneum*; and
R_{vt} is the diffusional resistance of the viable tissue.

This infers that the diffusional resistance across the *stratum corneum* has to be represented in terms of an aqueous pore pathway and a lipoidal pathway, which are in parallel and functionally independent. Thus, the total steady-state flux through the *stratum corneum*, J_{sc}, is the sum of two pathway contributions:

$$J_{sc} = J_{pore} + J_{lipid} = A\left(P_{pore} + P_{lipid}\right)\Delta C \tag{3.22}$$

where J_{pore} and J_{lipid}, and P_{pore} and P_{lipid} refer, respectively, to the aqueous pore pathway and the intercellular lipid routes.

The total resistance of the skin barrier as modelled is now represented by the expression:

$$R_T = \frac{1}{P_{pore} + P_{lipid}} + R_{sc} \tag{3.23}$$

where R_{sc}, defined in 3.21, is now expressed in terms of the contributing permeability coefficients of its parallel—and independent—diffusive routes.

Thus, the total diffusional resistance of the *stratum corneum* is the reciprocal sum of its contributing permeability coefficients, and individual resistances are the reciprocals of the individual permeability coefficients. This may be summarised in an expression that contains all contributing factors:

$$R_T = \frac{1}{\frac{1}{R_{pore}} + \frac{1}{R_{lipid}}} + R_{vt} \tag{3.24}$$

The diffusional resistances are now replaced with their equivalents in terms of fractional areas:

$$R_T = \frac{1}{\frac{f_{pore} \cdot D_{pore}}{h_{pore}} + \frac{f_{lipid} \cdot K_{lipid/water} \cdot D_{lipid}}{h_{lipid}}} + \frac{k_{vt}}{D_{vt}} \tag{3.25}$$

where

f is the diffusion by parallel paths;
D is the diffusion coefficient;
h is the thickness of the diffusing layer; and
K is the partition coefficient.

Equation 3.25 may be rewritten as:

$$R_T = \frac{h_{pore}.h_{lipid}}{f_{pore} \cdot D_{pore} \cdot h_{lipid} + f_{lipid} \cdot K_{lipid/water} \cdot D_{lipid} \cdot h_{pore}} + \frac{k_{vt}}{D_{vt}} \tag{3.26}$$

It is assumed in Eqs. 3.25 and 3.26 that the penetrant (or, As Flynn describes it, the pollutant) should be water soluble and that the partition coefficients for the aqueous pore path and the viable tissue path should have values which approach unity. Thus, for Eq. 3.26, when $K_{lipid/water} < 1.0$, which represents the polar extreme the diffusional resistance of the skin can be expressed as:

$$R_T = \frac{h_{pore}}{f_{pore} \cdot D_{pore}} + \frac{h_{vt}}{D_{vt}} \tag{3.27}$$

Also, as $f_{pore}.D_{pore} < D_{vt} \frac{h_{vt}}{D_{vt}}$ is negligible Eq. 3.27 becomes:

$$R_{\mathrm{T}} = \frac{h_{\mathrm{pore}}}{f_{\mathrm{pore}} \cdot D_{\mathrm{pore}}} \tag{3.28}$$

Equation 3.28 is defined by Flynn as the reciprocal of the operative permeability coefficient. It suggests that highly polar substances should partition independently. In such circumstances, the steady-state flux was defined:

$$J_{\mathrm{T}} = A\left(\frac{f_{\mathrm{pore}}.D_{\mathrm{pore}}}{h_{\mathrm{pore}}}\right)\Delta C \tag{3.29}$$

When compounds of increasing lipophilicity are considered, a point is reached where $f_{\mathrm{lipid}}.K_{\mathrm{lipid/water}}.D_{\mathrm{lipid}} > f_{\mathrm{pore}}.D_{\mathrm{pore}}.h_{\mathrm{lipid}}$. Beyond this point, the following expression may be considered:

$$R_{\mathrm{T}} = \frac{h_{\mathrm{lipid}}}{f_{\mathrm{lipid}} \cdot K_{\mathrm{lipid/water}} \cdot D_{\mathrm{lipid}}} + \frac{h_{\mathrm{vt}}}{D_{\mathrm{vt}}} \tag{3.30}$$

At the point of transition considered by Flynn, $f_{\mathrm{lipid}}.K_{\mathrm{lipid/water}}.D_{\mathrm{lipid}} < D_{\mathrm{vt}}$, therefore:

$$R_{\mathrm{T}} = \frac{h_{\mathrm{lipid}}}{f_{\mathrm{lipid}} \cdot K_{\mathrm{lipid/water}} \cdot D_{\mathrm{lipid}}} \tag{3.31}$$

As it follows that the operative permeability coefficient is the reciprocal of the equation, then in a region of intermediate polarity, the following expression applies:

$$J_{\mathrm{T}} = A\left(\frac{f_{\mathrm{lipid}} \cdot D_{\mathrm{lipid}} K_{\mathrm{lipid/water}}}{h_{\mathrm{lipid}}}\right)\Delta C \tag{3.32}$$

Equations 3.31 and 3.32 would suggest that there is a second area where permeability is directly dependent on the partition coefficient. Flynn suggest that such a region covers most, but not all, of the organic compounds of interest while commenting that some of the most persistent pollutants are some of the least polar compounds with which he was concerned. Thus, in approaching the non-polar extreme, the partition coefficient of a potential permeant is increased to the point where diffusional resistance of the viable tissues disappears, leaving the expressions:

$$R_{\mathrm{T}} = \frac{h_{\mathrm{vt}}}{D_{\mathrm{vt}}} \tag{3.33}$$

and

Table 3.1 Algorithms for calculating permeability coefficient (k_p) (modified from Flynn 1990)

	Low molecular weight compounds (<150 Da)	High molecular weight compounds (>150 Da)
Log $K_{ow} < 0.5$	Log $k_p = -3$	Log $k_p = -5$
$0.5 \leq$ Log $K_{ow} \leq 3.0$	Log $k_p = \log K_{ow} - 3.5$	
$0.5 \leq$ Log $K_{ow} \leq 3.5$		Log $k_p = \log K_{ow} - 5.5$
Log $K_{ow} > 3.0$	Log $k_p < -0.5$	
Log $K_{ow} > 3.5$		Log $k_p = -1.5$

$$J_{\mathrm{T}} = A\left(\frac{D_{\mathrm{vt}}}{h_{\mathrm{vt}}}\right)\Delta C \tag{3.34}$$

Equation 3.34 suggests that permeability should level off and become independent of partitioning as potential permeants become increasingly lipophilic. Indeed, Flynn suggests that a decline in permeability might be expected. Thus, a compartmentalised classification of skin permeability was developed which is not global in its scope and which also suggests that the process of permeability—in the context of molecular physicochemical properties—is not linear.

Flynn then produced a series of models featuring sigmoidal plots which qualitatively fitted the general pattern of the data. These were generally classified into the groups shown in Table 3.1.

In the main part of the model—these are of intermediate lipophilicity—the permeability coefficients of low molecular weight chemicals increased exponentially, reflecting the lipophilic nature of the skin barrier expressed mathematically, above. Further, the polar molecules in the data set appear to show evidence of a polar pore pathway with permeability (from the literature examples in the data set) limited to approximately 6×10^{-4} cm/h. The molecules on the lipophilic extreme appear to "top out" at an approximate permeability of 3×10^{-1} cm/h.

Thus, Flynn produced the ability to estimate the permeability of a penetrant from its partition coefficient, and did so using a model that incorporates a system of classification based on lipophilicity and molecular weight. While later researchers have argued for a single representative—and, indeed, Flynn commented that his model was only an "interim tool" until a unified model was developed—it is clear that the theoretical basis of this approach, particularly the dependence of permeability on the physiochemical properties of a penetrant, laid significant foundations for the development of further quantitative studies. It also collated and made available a single data set from which researchers subsequently developed a large number and range of models of skin permeability.

Flynn, however, did more than simply publish a collated data set. In the original research by Flynn (1990), a number of algorithms to predict skin permeability were proposed. Flynn presented a simple approach with algorithms for low and high molecular weight compounds (see Table 3.1). The algorithms stated that very hydrophilic and hydrophobic compounds had low and high skin permeability, respectively, and that a range of lipophilicity-dependent QSARs could be used to predict skin permeability for the high and low molecular weight compounds. No statistical measures of fit were provided at this time.

El Tayar et al. (1991) took human skin permeation data from Flynn's data set; specifically, they analysed various subsets of Flynn's data set to determine quantitative relationships with physicochemical properties and structural descriptors of the penetrants. Their analyses suggested a significant role for hydrophobicity in mediating skin permeation for a number of the subsets they analysed, which was inversely correlated with lipophilicity. This was expressed by the term $\Delta\log P_{(\text{oct-hep})}$, which is a measure of the difference between the log P_{octanol} and log P_{heptane}, which the authors described as being mainly a measure of the H-bond donor activity (hydrogen bond donor acidity) of the solutes. Further, they suggested that log P_{octanol} in some cases correlates positively to skin permeation. They also found that no correlations existed between the molecular weights of permeants and solvent-accessible surface areas. They interpreted their results in the context of a "unified mechanism model" which suggested that some penetrants permeated the skin via the intercellular route, which correlated with both $\Delta\log P_{(\text{oct-hep})}$ and log P_{heptane}, and that others penetrated the skin via a transcellular route, which was correlated with log P_{octanol} only. The significance of hydrogen bonding in the process of percutaneous absorption is an issue that will be discussed in subsequent chapters. Further, the role of molecular weight in the absorption process is one that was almost immediately addressed by Potts and Guy.

Potts and Guy (1992) effectively quantified the approach taken by Flynn (1990), and their work has essentially laid the foundations of a whole new area of skin permeability research. They took published human skin permeability coefficient data (as k_p, published by a number of researchers, taken from sixteen different published articles and collated by Flynn: Scheuplein 1965; Scheuplein et al. 1969; Scheuplein and Blank 1971; Michaelis et al. 1975; Roy and Flynn 1989, 1990; Hadgraft and Ridout 1987; Roberts et al. 1977, 1978; Dutkiewicz and Tyras 1967; Jolicoeur et al. 1992; Raykar et al. 1988; Anderson et al. 1988; Chowhan and Pritchard 1978; Shaw and Chandrasekaran 1978; Dutkiewicz and Tyras 1968) and analysed it using a simple model based on octanol/water partition coefficient and molecular size. They used a data set of 93 compounds which occupy a large molecular space, in terms of lipophilicity (log $K_{o/w}$ −3 to +6) and molecular weight (180–750), inferring the relevance of their single model to this range, based on predictive power and sitting somewhat contradictorily with Flynn's classification—of the same data in the same molecular space—of the data into a series of distinct models.

Citing Kasting's work (Kasting et al. 1987), which applies similar methods to a smaller data set they comment that, as 30 % variation in experimental permeability data is common, their model completely describes the data (Guy and Hadgraft 1988). They also comment that individual analyses of the larger "subsets" from which the Flynn data set was constructed produced models which were more accurate, where up to 82 % of the variation could be accounted for by the model. This also infers that as different subsets are collated, the quality of the model lessens. Potts and Guy also suggest that the use of molecular weight was generally equivalent in statistical quality to the use of molecular volume.

Therefore, their study demonstrated the use of lipophilicity in combination with either molecular weight or molecular volume to predict the skin permeabilities that Flynn had collated. Thus, for 93 compounds they reported the following relationship:

$$\log k_{\rm p}({\rm cm/s}) = -6.3 + 0.71 \log P - 0.0061\,{\rm MW} \tag{3.35}$$

$$\left[r^2 = 0.67\right]$$

where

$k_{\rm p}$ is the skin permeability coefficient;
$\log P$ often written as $\log K_{\rm o/w}$, is the octanol–water partition coefficient; and
MW is the molecular weight.

The experimental values used to derive this expression included chemicals with molecular weights ranging from 18 to over 750 and log *P* values from −3 to +6. This expression suggests that the rate-limiting step in the process of percutaneous absorption is the lipophilic *stratum corneum*. This suggests that, for lipophilic compounds, it will yield values of $k_{\rm p}$ that are unrealistically high and that the equation does not consider the influence of the underlying tissues of the viable epidermis and dermis on the prediction of permeability, which had been postulated previously (Michaels et al. 1975).

The "Potts and Guy" equation is a simple, powerful and widely applicable representation of the skin permeability process. Its simplicity is placed firmly in the context of Flynn's work, where the approximation of any model used to estimate "real-world" permeability is clear. Its use of relevant (to physical and life scientists), well understood and widely used physicochemical descriptors is central to the success of this model and provides a transparent algorithm which clearly expresses significant mechanistic insight into skin permeability—in this case, their model is consistent with Leib and Stein's comments on transport through lipid lamellae (Lieb and Stein 1971). The simplicity of this model extends to the absence of a model for the aqueous, or polar, pore pathway which was described by Flynn, among others. Potts and Guy comment that the perceived independence of log $k_{\rm p}$ upon $K_{\rm oct}$ for small, polar molecules (compared, for example, to the linear relationship observed in their data set between log $k_{\rm p}$ and log $K_{\rm oct}$ for larger, generally more lipophilic compounds) is explained by their molecular volumes, suggesting no need for a separate model to describe these compounds in such a diverse data set.

Further, they cite the example of poorly permeable members of the data set—sucrose and glucose—which are predicted reasonably by Eq. 3.35. They conclude by considering that the sigmoidal dependence of log $k_{\rm p}$ upon log $K_{\rm oct}$—in effect, its nonlinearity—should be considered in the context of molecular volume. In this context, they suggested that the data lie on a three-dimensional surface characterised by log k_p, log K_{oct} and molecular volume. They conclude that the consideration of aqueous pores is not necessary and that a single, simple algorithm can describe skin permeability with excellent predictive ability ($r^2 = 0.67$).

Thus, the "Potts and Guy" equation, as it is now widely known (despite subsequent publication by the same authors of other algorithms describing skin permeability!), provided a simple model which describes the process of skin permeability in terms of lipophilicity and molecular weight. This gives it a mechanistic simplicity, but also a relevance and value, to scientists studying skin permeability and who may not necessarily be familiar with more complex molecular descriptors which are often specific to fields such as computational chemistry, and which themselves may lack a suitable context with regard to skin permeability. Potts and Guy claim excellent predictive power for their equation, and this has provided enormous value to formulation scientists and toxicologists since the publication of the model.

References

Albery WJ, Hadgraft J (1979) Percutaneous absorption: theoretical description. J Pharm Pharmacol 31:129–139

Anderson BD, Higuchi WI, Raykar PV (1988) Heterogeneity effects on permeability-partition coefficient relationships in human stratum corneum. Pharm Res 5:566–573

Aulton ME (2003) Pharmaceutics: the science of dosage form design, 3rd edn. Churchill Livingstone, London

Barry BW (1983) Dermatological Formulations: percutaneous Absorption. Marcel Dekker, New York

Berthelot M, Jungfleisch E (1872) On the laws that operate for the partition of a substance between two solvents. Ann Chim Phys 26:396–407

Blank IH (1969) Transport across the *stratum corneum.* Toxicol Appl Pharmacol 3:23–29

Chandrasekaran SK, Bayne W, Shaw JF (1978) Pharmacokinetics of drug permeation through human skin. J Pharm Sci 67:1370–1374

Chowhan Z, Pritchard R (1978) Effect of surfactants on percutaneous absorption of naproxen. 1. Comparisons of rabbit, rat and human excised skin. J Pharm Sci 67:1272–1274

Cooper ER (1976) Pharmacokinetics of skinpenetration. J Pharm Sci 65:1396–1397

Crank J (1975) The mathematics of diffusion, 2nd edn. Clarendon Press, Oxford

Crum-Brown A, Fraser TR (1868) On the connection between chemical constitution and physiological action. I. On the physiological action of the salts of the ammonium bases, derived from strychnia, brucia, thebaia, codeia, morphia, and nicotia. Trans R Soc Edinb 25:151–203

Dearden JC (1985) Partitioning and lipophilicity in quantitative structure-activity relationships. Environ Health Perspect 61:203–228

Dutkiewicz T, Tyras H (1967) A study of the skin absorption of ethylbenzene in man. Br J Ind Med 24:330–332

Dutkiewicz T, Tyras H (1968) Skin absorption of toluene, styrene and xylene by man. Br J Ind Med 25:243

El Tayar N, Tsai RS, Testa B, Carrupt PA, Hansch C, Leo A (1991) Percutaneous penetration of drugs—a quantitative structure-permeability releationship study. J Pharm Sci 80:744–749

Ferguson J (1939) The use of chemical potentials as indices of toxicity. Proc Roy Soc London B127:387–403
Flynn GL (1990) Physicochemical determinants of skin absorption. In: Gerrity TR, Henry CJ (eds) Principles of route-to-route extrapolation for risk assessment. Elsevier, New York, pp 93–127
Franz TJ (1975) Percutaneous absorption: on the relevance of *in vitro* data. J Invest Dermatol 64:190–195
Guy RH, Hadgraft J (1982) Percutaneous metabolism with saturable enzyme kinetics. Int J Pharm 11:187–197
Guy RH, Hadgraft J (1984) Pharmacokinetics of percutaneous absorption with concurrent metabolism. Int J Pharm 20:43–51
Guy RH, Hadgraft J (1988) Physicochemical aspects of percutaneous penetration and its enhancement. Pharm Res 5:753–758
Guy RH, Hadgraft J (1983) Physicochemical interpretation of the pharmacokinetics of percutaneous absorption. J Pharmacokinet Biopharm 11:189–203
Guy RH, Hadgraft J, Maibach HI (1983) Percutaneous absorption: multidose pharmacokinetics. Int J Pharm 17:23–28
Hadgraft J, Ridout G (1987) Development of model membranes for percutaneous absorption measurements. 1. Isopropyl myristate. Int J Pharm 39:149–156
Hadgraft JW, Somers GF (1956) Percutaneous absorption. J Pharm Pharmacol 8:625–634
Higuchi T (1960) Physical chemical analysis of percutaneous absorption process from creams and ointments. J Soc Cosmet Chem 11:85–97
Higuchi T (1961) Rate of release of medicaments from ointment bases containing drugs in suspension. J Pharm Sci 50:874–875
Jolicoeur LM, Nassiri MR, Shipman C, Chio HK, Flynn GL (1992) Etorphine is an opiate analgesic physicochemically suited to transdermal delivery. Pharm Res 9:963–965
Kasting GB, Filloon TG, Meredith MP (1992) A method for improving the sensitivity of in vitro skin penetration studies. Pharm Res 9:S195
Kasting GB, Smith RL, Cooper ER (1987) Effect of lipid solubility and molecular size on percutaneous absorption. In: Shroot B, Schaefer H (eds) Skin pharmacokinetics. Karger, Basel, pp 138–153
Kroes R, Renwick AG, Feron V, Galli CL, Gibney M, Greim H, et al (2007) Application of the threshold of toxicological concern (TTC) to the safety evaluation of cosmetic ingredients. Food Chem Toxicol 45:2533–2562
Lieb WR, Stein WD (1971) Implications of two different types of diffusion for biological membranes. Nature 243:219–222
Magnusson BM, Anissimov YG, Cross SE, Roberts MS (2004) Molecular size as the main determinant of solute maximum flux across the skin. J Invest Dermatol 122:993–999
Martin A, Swarbrick J, Cammarata A (1983) Physical pharmacy, 3rd edn. Lea and Febiger, New York
Meyer H (1899) On the theory of alcohol narcosis I. Which property of anesthetics gives them their narcotic activity? Arch Exper Pathol Pharmakol 42:109–118
Michaelis AS, Chandrasekaran SK, Shaw JE (1975) Drug permeation through human skin: theory and *in vitro* experimental measurement. AIChE 21:985–996
Overton E (1897) Osmotic properties of cells in the bearing on toxicology and pharmacology. Z Physik Chem 22:189–209
Potts RO, Guy RH (1992) Predicting skin permeability. Pharm Res 9:663–669
Raykar PV, Fung MC, Anderson BD (1988) The role of protein and lipid domains in the uptake of solutes by human stratum corneum. Pharm Res 5:140–150
Richardson BW (1869) Physiological research on alcohols. Med Times Gazzette 2:703–706
Richet C (1893) On the relationship between the toxicity and the physical properties of substances. Compt. Rendus Seances Soc Biol 9:775–776
Riegelman S (1974) Pharmacokinetics. Pharmacokinetic factors affecting epidermal penetration and percutaneous absorption. Clin Pharmacol Ther 16:873–883

Roberts MS, Anderson RA, Swarbrick J, Moore DE (1978) Percutaneous absorption of phenolic compounds—mechamism of diffusion across stratum corneum. J Pharm Pharmacol 30:486–490

Roberts MS, Anderson RA, Swarbrick J (1977) Permeability of human epidermis to phenolic compounds. J Pharm Pharmacol 29:677–683

Roberts MS, Anissimov YG, Gonsalvez RA (1999) Mathematical models in percutaneous absorption. In: Bronaugh RL, Maibach HI (eds) *Percutaneous Absorption*, 3rd edn. Marcel Dekker, Inc. CRC Press, New York pp 3–56

Rothman S (1943) The principles of percutaneous absorption. J Lab Clin Med 28:1305–1321

Roy SD, Flynn GL (1989) Transdermal delivery of narcotic analgesics—comparative permeabilities of narcotic analgesics through human cadaver skin. Pharm Res 6:825–832

Roy SD, Flynn GL (1990) Transdermal delivery of narcotic analgesics—pH, anatomical and subject influences on cutaneous permeability of fentanyl and sufentanyl. Pharm Res 7:842–847

Scheuplein RJ, Blank IH, Brauner GI, MacFarlane DJ (1969) Percutaneous absorption of steroids. J Invest Dermatol 52:63–70

Scheuplein RJ, Blank IH (1971) Permeability of the skin. Physiol Rev 51:702–747

Scheuplein RJ (1965) Mechanism of percutaneous absorption. I. Routes of penetration and the influence of solubility. J Invest Dermatol 45:334–346

Shaw JE, Chandrasekaran SK (1978) Controlled delivery of drugs for systemic action. Drug Metab Rev 8:223–233

Sinko PJ (2010) Martin's physical pharmacy and pharmaceutical sciences. Lippincott, Williams & Wilkins, Philadelphia

Tojo K (1987) Random brick model for drug transport across *stratum corneum*. J Pharm Sci 76:889–891

Wallace SM, Barnett G (1978) Pharmacokinetic analysis of percutaneous absorption: evidence of parallel penetration pathways for methotrexate. J Pharmacokinet Biopharm 6:315–325

Walters KA, Roberts MS, Gettings S (1998) Percutaneous absorption of sunscreens. In: Bronaugh R, Maibach HI (eds) Percutaneous absorption, 3rd edn. Marcel Dekker, New York

Williams AC (2003) Transdermal and topical drug delivery. The Pharmaceutical Press, London

Woolfson AD, McCafferty DF (1993) Percutaneous local anaesthesia. Ellis Horwood, London

Chapter 4
The New Breadth of Research in the Field

At around the same time Potts and Guy were developing their global model of skin permeability based on Flynn's data set (Flynn 1990; Potts and Guy 1992), other researchers were also exploring similar research themes. For example, El Tayar et al. (1991) examined a data set of skin permeability for chemicals taken from the literature. In doing so they confirmed the role of lipophilicity as a major significant parameter in skin permeability in most, but—interestingly—not all, cases. They further described, indirectly, a significant correlation to hydrogen bond donor acidity of solutes using the partition coefficient of heptane/water subtracted from the octanol–water partition coefficient ($\Delta \log K_{\text{oct-hep}}$). They did not find, as Potts and Guy did, any significant role for molecular size in the mechanism of skin permeability. The issue of hydrogen bonding and its apparent role in skin permeability will be discussed in detail below.

Pugh and Hadgraft (1994) used a novel, so-called ab initio approach to model skin permeability. This was a fragment method based on various molecular substructures within each chemical. Their analysis, using a linear model, of the skin permeability of a 91-member data set collated from the literature, against a range of common molecular descriptors, resulted in a series of complex 17- and 11-descriptor models which were in general agreement with, but not significantly better than, the Potts and Guy equation. Moss et al. (2002) commented that Pugh and Hadgraft's fundamental empirical approach to addressing the issue of predicting skin permeability lacked the uniform approach, mechanistic simplicity and utility of the Potts and Guy equation, a consideration which, in hindsight, appears to be borne out by their comparative widespread use (as suggested by the relative citations of each study in the literature) but which also, more importantly, reflects the importance of a simple and transparent model—and approach to modelling—in developing our understanding of the processes of skin permeability. However, Pugh and Hadgraft's approach resulted in the identification of a number of outliers in the data set (including atropine, estriol, naproxen, nicotine and toluene) which were confirmed by subsequent studies. Interestingly, they also commented that, where predictions differed significantly from experimental values, this might indicate compounds whose permeability process is in some way abnormal or does not fit the general model.

G.P. Moss et al., *Predictive Methods in Percutaneous Absorption*,
DOI 10.1007/978-3-662-47371-9_4

Lien and Gao (1995) examined a subset of 22 chemicals from the Flynn data set. They developed an algorithm relating the permeability coefficient to the logarithm of molecular weight and to a biphasic response to hydrophobicity. In addition, they also found that the number of hydrogen bonds that may be formed by a compound (H_b) was significant in modelling skin permeability (k_p, as cm/h):

$$\log k_p = 0.84 \log K_{ow} - 0.07(\log K_{ow})^2 - 0.27H_b - 1.84 \log MW + 4.39$$
$$[n = 22 \quad r^2 = 0.96 \quad s = 0.30 \quad F = 93.6] \tag{4.1}$$

Clearly, this model is limited by the size of the data set used. Nevertheless, this model is significant and interesting as it introduces nonlinearity (as log K_{ow}^2) into the relationship between skin permeability and physicochemical parameters. It is also unique in that it presents log MW rather than MW as a significant parameter. Lien and Gao also commented that the correlations obtained were dependent not only on the biological system but also on the vehicle used. Indeed, in the absence of a lipophilic vehicle they determined that the ideal lipophilicity for optimal human skin permeability was achieved by a potential permeant with a log*P* value ranging from 2.5 to 6, the higher end of this range being an extrapolation based on the molecular space of their data set. The use of a lipophilic vehicle (in model studies featuring mouse skin as a membrane) saw this ideal range drop to approximately 0.4, potentially due to changes in penetrant solubility in the vehicle or potential effects of the solvent on the integrity of the *stratum corneum* barrier. They suggest that, while an increase in MW was characteristic of a decrease in permeability, increasing hydrogen bonding potential could have either a slightly positive or slightly negative effect.

In a major development of, and from, the Potts and Guy approach, Cleek and Bunge (1993) described a range of models that estimated dermal absorption under a range of conditions. They commented that Potts and Guy's (1992) model, given the inherent variation associated with skin permeability data, is a robust model which has been tested rigorously by the nature of its input data. The Potts and Guy equation averages most of the experimental variability and, as such, the weight of the experimental data strongly supports the validity of the correlation prediction. Thus, data which deviate significantly from the Potts and Guy equation (Eq. 3.36) should be evaluated with regard to the experimental procedure used in generating the data and data reproducibility before these data are accepted as being more representative than the correlation. Potts and Guy (1992) suggested that data varying from Eq. 3.36 by a factor of three or more would lie outside the 95 % confidence interval. These comments, from both Potts and Guy and Cleek and Bunge, imply that the equation can fit the data and that any erroneous data which are a poor fit are due to a poor experimental design and/or execution and not due to limitations of, or deficiencies in Eq. 3.36.

Cleek and Bunge went on to develop models based on three experimentally relevant scenarios: a single finite membrane, a semi-finite membrane and a finite two-membrane composite. All models have been, in various forms, reported previously

(Crank 1975, Carslaw and Jaeger 1980; Ozisik 1980). All these models assumed that the membranes—either the *stratum corneum*, or the *stratum corneum* considered with the viable epidermis—are passive with regard to diffusivity, thickness and partition coefficients; that is, the vehicle and the absorbing chemical do not alter the nature or composition of the membrane. Cleek and Bunge consider the heterogeneous epidermal membranes as mathematically pseudo-homogeneous, resulting in the diffusivity, thickness and partition coefficient values as being effective properties. Thus, the *stratum corneum* can be considered as a single finite membrane:

$$\frac{M_{\text{in}}}{AL_cK_{cv}C_v^0} = \tau + \frac{1}{3} - \frac{2}{\pi^2}\sum_{n=1}^{\infty}\frac{e^{-n^2\pi^2\tau^2}}{n^2} \tag{4.2}$$

where

A	is the surface area of chemical exposure
L_c	is the thickness of the *stratum corneum*
K_{cv}	is the equilibrium partition coefficient between the *stratum corneum* and the vehicle of the absorbing chemical
C_v^0	is the concentration of the absorbing chemical in the vehicle/ formulation, which is assumed to remain constant during the exposure time, t_{exp}
$\frac{M_{\text{in}}}{AL_cK_{cv}C_v^0}$	is the normalised cumulative mass absorbed into the *stratum corneum*
$\tau = \frac{t_{\text{exp}}D_c}{L_c^2}$,	with t_{exp} being the exposure time normalised by the characteristic diffusion time $(\frac{D_c}{L_c^2})$ for a chemical in the *stratum corneum*
D_c	is the effective diffusivity of the absorbing chemical in the *stratum corneum*
n	is the summation index (Cleek and Bunge 1993).

Thus, the normalised cumulative mass out of the *stratum corneum* is calculated by integrating the flux at L_c over the exposure time:

$$\frac{M_{\text{out}}}{AL_cK_{cv}C_v^0} = \tau - \frac{1}{6} - \frac{2}{\pi^2}\sum_{n=1}^{\infty}\frac{(-1)^n e^{-n^2\pi^2\tau^2}}{n^2} \tag{4.3}$$

where M_{out} is the normalised cumulative mass leaving the *stratum corneum*.

Removal of the infinite series from 4.2 and 4.3 makes the equations more widely practical:

$$\frac{M_{\text{in}}}{AL_cK_{cv}C_v^0} = \tau + \frac{1}{3} \tag{4.4}$$

$$\frac{M_{\text{out}}}{AL_cK_{cv}C_v^0} = \tau - \frac{1}{6} \tag{4.5}$$

Cleek and Bunge commented that chemicals with different K_{cv} values, indicated by different K_{ow} values, can still exhibit similar lag times as long as their D_c values were similar, suggesting that lag time is dependent on diffusivity and that lag time and diffusivity may not be directly related to log K_{ow}.

They then proposed a semi-infinite membrane model, which was applied to the situation where the exposure time was short and where the penetrant will therefore only penetrate a short distance into the *stratum corneum*. In such cases, the *stratum corneum* is considered to be comparatively thick, or semi-infinite. In such cases, the normalised cumulative mass is given by:

$$\frac{M_{\text{in}}}{AL_cK_{cv}C_v^0} = 2\sqrt{\frac{D_c t_{\text{exp}}}{L_c^2 \pi}} = 2\sqrt{\frac{\tau}{\pi}} \tag{4.6}$$

They considered a finite two-membrane composite model which includes both the *stratum corneum* and viable epidermal layers, which provides a more realistic model to skin permeation than single-membrane systems:

$$\begin{aligned} \frac{M_{\text{in}}}{AL_cK_{cv}C_v^0} = {} & \frac{1}{1+B}\left[\tau + \frac{G(1+3B)+B(1+3BG)}{3G(1+B)}\right] \\ & + 2(1+B)\sum_{n=1}^{\infty}\frac{\sin\left(\lambda_n/\sqrt{G}\right)\sin\lambda_n\exp\left(-\lambda_n^2\tau\right)}{\lambda_n^2\sigma_n} \\ & - 2(1+B)\sum_{n=1}^{\infty}\frac{\cos\left(\lambda_n/\sqrt{G}\right)\cos\lambda_n\exp\left(-\lambda_n^2\tau\right)}{B\sqrt{G}\lambda_n^2\sigma_n} \end{aligned} \tag{4.7}$$

where B is the permeability ratio between the *stratum corneum* and epidermis and measures the relative contribution of the *stratum corneum* and viable epidermis resistances

$$\sigma_n = \frac{1}{BG}\left[\sqrt{G}(1+B)\cos\left(\lambda_n/\sqrt{G}\right)\cos(\lambda_n) - (1+GB)\sin\left(\frac{\lambda_n}{\sqrt{G}}\right)\sin\lambda_n\right]$$

and λ_n, which are both eigenvalues.

$G = \frac{L_c^2 D_e}{L_e^2 D_c}$, the ratio of the lag times for the *stratum corneum* and viable epidermis.

Equations 4.3 and 4.6 yield identical results during shorter exposure times as the chemical does not penetrate far enough into the *stratum corneum* to make the finite nature of this layer relevant. Equation 4.6 therefore describes absorption into the *stratum corneum* during the non-steady-state period of the diffusion process.

Steady-state mass absorption is determined from a simplified version of Eq. 4.7:

$$\frac{M_{\text{in}}}{AL_cK_{cv}C_v^0} = \frac{1}{1+B}\left[\tau + \frac{1+3B+3B^2}{3(1+B)}\right] \tag{4.8}$$

This means the diffusion process no longer depends on *G*, and Cleek and Bunge commented that this expression is accurate for large (>10) values of G which reflects the lack of a compromised *stratum corneum* barrier. Further, in the context of the partition between the viable epidermis and the *stratum corneum*, and the role of the former in influencing skin permeability, Cleek and Bunge also derived an expression for k_p that sets an upper limit for permeation of highly lipophilic species:

$$k_p^{\text{adj}} = \frac{k_p}{1 + \left(1400 \cdot k_p \cdot \sqrt{\text{MW}}\right)} \tag{4.9}$$

where k_p is the predicted permeability coefficient calculated from a QSAR equation, such as that presented by Potts and Guy (1992).

The overall power of this study is that it includes the effects of chemical characteristics (i.e. log *P* and molecular weight) and exposure—particularly exposure time—into models of permeant uptake in skin in both steady-state and non-steady-state conditions. It also considers the skin "compartments" and provides a more balanced model that considers the molecular requirements for permeability in the context of partitioning between the skin (epidermal) layers. Thus, the cumulative mass absorbed into the *stratum corneum* can be used to assess dermal exposure risks.

Bunge, Cleek and co-workers developed the above study with two subsequent publications. They next examined the effect of molecular weight and lipophilicity on dermal absorption (Bunge and Cleek 1995). They emphasised the influence of partitioning between the *stratum corneum* and the viable epidermis (defined as *B*) in the context of key physicochemical properties and considered the balance of properties required for effective skin permeation rather than simply considering partitioning into the lipid *stratum corneum.* They considered four main models which could be used to estimate *B*. These models demonstrated, for example, how the value of *B* varied with log K_{ow} across the four models for a molecule with a constant molecular weight (300). Their methods vary in how they consider the hydrophilic–hydrophobic balance between the viable epidermis and *stratum corneum* and how this manifests itself in the determination of *B*. Their first method considers that the hydrophilic–hydrophobic balance between the viable epidermis and *stratum corneum* varies according to log K_{ow}, specifically that $B = \frac{K_{ce}D_cL_e}{D_eL_c} = \frac{K_{\text{ow}}}{1150}$, which assumes that molecular weight dependence is the same in both the viable epidermis and the *stratum corneum.* Correcting this estimate for steady-state permeability of the *stratum corneum* suggests that the effect attributed to K_{ow} by Model 1 is too strong. Model 2 makes a similar assumption but considers *B* to be a function of $K_{\text{ow}}^{0.74}/230$ and also suggests that *B* does not change with molecular weight. Model three suggests that molecular weight dependence is exponential for the *stratum corneum* and that there is no molecular weight dependence for the viable epidermis, and is represented by $B = \frac{P_{cw}}{0.36(\text{cm/h})}$. Model 4 assumes that B is dependent exponentially on molecular weight for the *stratum corneum* and inversely related to the square root of molecular weight for the viable epidermis, being represented by $B = \frac{P_{cw}\sqrt{\text{MW}}}{2.6(\text{cm/h})}$. Models 3 and 4 predict that B decreases as molecular weight increases. Model 3 predicts a more rapid decrease in B, compared to Model 4, as it assumes that permeability in the viable epidermis is unaffected by molecular weight, whereas *stratum corneum* permeability decreases exponentially. Thus, with the caveat of considering very small molecules (i.e. MW < 36), they suggest that their fourth model is the most realistic, as it suggests a more moderate decrease in permeability across the viable epidermis decreases with increasing molecular weight but more moderately than the permeability across the *stratum corneum*. Thus, in most cases, estimates of cumulative mass absorbed can be made from D_c, L_c and K_{cv} (as defined above) and B, the relative size of *stratum corneum* permeability to the viable epidermis permeability. Their modelling studies suggested that B is only important for highly lipophilic, small molecular weight permeants. When B is considered important Bunge and Cleek recommend the application of the Potts and Guy (1992) algorithm to determine skin permeability.

Finally, in this series, Bunge et al. (1995) compared models of steady-state and non-steady-state absorption for estimating dermal absorption. The non-steady-state methods were considered in the cases of short exposure times and larger absorption, as well as considering the contribution of the hydrophilic barrier of the viable epidermis to the passage of lipophilic chemicals. They found that, for example cases, the steady-state modelling approach significantly underestimated the dermal absorption under non-steady-state conditions, meaning that permeability values calculated from data sets which include non-steady-state data will be incorrect. They further determined that calculating permeability values from cumulative absorption data measured for exposure times less than 18 times the *stratum corneum* lag time overestimates permeability. Models of non-steady-state diffusion, including those in the context of finite dosing, are discussed in greater detail in Chaps. 6 and 8.

Barratt (1995) also analysed the data set published by Flynn (1990), using the log P and log of permeability coefficients (listed by Barratt as log *PC*) published by Flynn. He calculated the molecular volumes for each member of the data set and obtained melting point values from literature sources. A series of cross-validated QSARs were derived from these data, and the data were also subjected to principal component analysis. The chemicals in the data set were categorised as (a) steroids, (b) other pharmacologically active chemicals and (c) small organic molecules without pharmacological activity. In categorising the data in this way he noted that the subset of hydrocortisone derivatives were consistently modelled poorly by these methods (this is an issue that will be discussed in detail later). QSARs were derived for each of these subdivisions of Flynn's data set.

Barratt's use of melting point is not explained in detail, other than to comment on its use as a parameter that reflects solubility. It is difficult to elaborate on the significance of the use of this parameter in this study. Melting point, however, is dependent on hydrogen bonding and as such may infer indirectly the importance of that phenomenon in skin permeation. Its inclusion resulted in a slight increase in the amount of variance explained by the model (compared to Potts and Guy (1992)) from 71.1 to 76.6 %. In the context of Potts and Guy's comments on the variability of data (that up to 30 % variance is common in skin permeability experiments) and the additional context that Flynn's data set is constructed from a number of different literature sources and therefore a number of different laboratory studies, Barratt concluded that models of this nature could describe fully the variance in the data.

Barratt further commented that principal component analysis demonstrated the existence of distinct groupings within the data—this, again, is interesting in the context of subsequent Gaussian process studies of skin absorption, which will be discussed later, as well as the issues raised by Johnson et al. (1995), and later quantified by Moss and Cronin (2002) regarding the quality of some of the original data from the Flynn data set. The most significant QSAR reported by Barratt (units cm/h) was derived from a subset of 60 "small molecules and steroids", but excluding the hydrocortisone derivatives, taken from Flynn's data set:

$$\begin{aligned} &\log k_p = 0.82 \log K_{\mathrm{ow}} - 0.0093\,\mathrm{MV} \ - 0.039\,\mathrm{MPt} \ - 2.36 \\ &\left[n = 60 \quad r^2 = 0.90 \quad s = 0.39 \quad F = 176\right] \end{aligned} \tag{4.10}$$

The hydrocortisone derivatives were poorly modelled by this expression and this was attributed to their analysis within the same laboratory environment (Anderson et al. 1988; Raykar et al. 1988) and, by Barratt, to a particular experimental technique associated with this laboratory. Similar conclusions were drawn for the underestimation of permeability by the model for ethylbenzene, styrene and toluene. Despite the use of a linear model Barratt is clear that the overall trend, extrapolated beyond the molecular space of this data set, is not linear as increases in lipophilicity will increase permeability only to the point where aqueous solubility in skin layers beneath the *stratum corneum* permits further permeation. Barratt's final conclusion is that while the model explains 90 % of the variance in the data and is expected to yield accurate predictions, it has not been tested experimentally.

Another study around the same time—in what could be described as another analysis of subsets of Flynn's data set—was reported by Potts and Guy (1995). They investigated the role of hydrogen bonding by developing a predictive model based on a subset of 37 non-electrolytes. The data they used was again taken from Flynn's data set. The resulting QSAR highlighted the significance of hydrogen bonding but did not feature a hydrophobicity term. The QSAR derived was as follows:

$$\begin{aligned} &\log k_p(\mathrm{cm/s}) = 0.0256\,\mathrm{MW} - 1.72 \sum \alpha_2^{\mathrm{H}} - 3.93 \sum \beta_2^{\mathrm{H}} - 4.85 \\ &\left[n = 37 \quad r^2 = 0.94 \quad s \text{ not given} \quad F = 165\right] \end{aligned} \tag{4.11}$$

where
MV is molecular volume
$\sum \alpha_2^{\mathrm{H}}$ is the solute hydrogen bond acidity
$\sum \beta_2^{\mathrm{H}}$ is the solute hydrogen bond basicity.

Equation 4.11 is interesting in that it highlights a significant issue with the QSAR (and related) analyses of skin permeability. By altering the input to the model (the membership of the data set, including its size), it substantially alters the output—the QSAR algorithm. This study, and its QSAR, indicates that the factors (molecular descriptors) influencing the skin permeability of non-electrolytes are significantly different to those which influence permeability for the whole data set. In particular, it provides an understanding that hydrogen bond activity is inversely related to skin permeability and that the hydrogen bond acceptor ability of a potential permeant appears to be more significant than its hydrogen bond donor ability, linking potential permeation directly to permeant structure. However, a direct comparison with Potts and Guy (1992) is not possible as that model focused on a model of permeability based on lipophilicity and molecular volume/weight, and did not specifically explore effects of hydrogen bonding on permeability. Further, the reduced size of the data set lessens the power of the model as a viable predictive model compared to their earlier study.

At a similar time, a substantial analysis, resulting in a range of QSAR models, of skin permeability was published by Abraham et al. (1995). Their study was based on earlier work by this group (Abraham 1993) and relates to their application of a general solvation equation:

$$\log \mathrm{SP} = c + rR_2 + s\pi_2^{\mathrm{H}} + \sum \alpha_2^{\mathrm{H}} + b \sum \alpha\beta_2^{\mathrm{H}} + vV_x \tag{4.12}$$

where
SP is the property of a series of solutes in a system; in the context of skin permeability, this is most likely k_p, but could also be steady-state flux.
R_2 is an excess molar refraction and is determined from knowledge of the compound refractive index, according to an earlier method by this group (Abraham et al. 1990). This term describes the tendency of a solute to interact with a phase through π or n electron pairs.
$s\pi_2^{\mathrm{H}}$ is the solute dipolarity/polarisability, defined by Abraham et al. (1991).
$\sum \alpha_2^{\mathrm{H}}$ represents solute-effective, or overall hydrogen bond acidity
$\sum \beta_2^{\mathrm{H}}$ represents solute-effective, or overall hydrogen bond basicity
V_x is the McGowan characteristic volume.

Their data sets were examined using multiple linear regression analysis. A series of subsets of the larger data set were examined. For example, they examined the data subsets which were previously investigated by El Tayar et al. (1991). Abraham reports that “very poor” correlations were often obtained in this analysis, attributing this to the values of alcohols and steroids lying on two distinct lines in a plot of log P_{oct} against −log k_p, although the utilisation of small data sets and the extrapolation of results from them might also contribute to poor models with limited statistical value. Further, El Tayar commented that the poor correlations were due to the Δlog P term, which reflects solute hydrogen bond acidity, which has a retarding effect on skin permeability due to hydrogen bond formation between permeants and hydroxyl groups in the alkyl side chains of ceramides during permeation via the intercellular route. Abraham and co-workers proposed an alternative equation to that presented by El Tayar, using the same data but based on their generalised model:

$$\log k_p\,(\text{cm/s}) = -5.333 - 0.622\pi_2^{\text{H}} - 0.378\sum\alpha_2^{\text{H}} - 3.342\sum\beta_2^{\text{H}} + 1.851V_x$$
$$[n = 22 \quad p = 0.9781 \quad \text{SD} = 0.260 \quad F = 93.7] \tag{4.13}$$

They then added three further permeability data—values for diethyl ether, 2-ethoxyethanol and butanone, taken from Scheuplein and Blank (1971), which resulted in the following expression:

$$\log k_p(\text{cm/s}) = -5.3194 - 0.567\pi_2^{\text{H}} - 0.506\sum\alpha_2^{\text{H}} - 3.368\sum\beta_2^{\text{H}} + 1.767V_x$$
$$[n = 25 \quad p = 0.9780 \quad \text{SD} = 0.260 \quad F = 110.1] \tag{4.14}$$

Thus, the models of Abraham et al. are significantly different in their mechanistic insight than those produced by El Tayar et al., with the former study focusing on the role of hydrogen bonding and polarisability, which means it sits close to Potts and Guy's algorithm for skin permeability (Potts and Guy 1992). Similar findings were, however, found by Abraham and colleagues for other subsets of the Flynn data set, including analysis of the Roberts et al. (1977) subset, where Abraham did not find evidence to support Roberts' proposal that phenols permeated the skin by two different routes based on their lipophilicities. They showed that, for phenols, there is a significant contribution from solute acidity which was not relevant for alcohols and steroids:

$$\log k_p(\text{cm/s}) = -4.994 - 0.341\pi_2^{\text{H}} - 1.691\sum\alpha_2^{\text{H}} - 2.689\sum\beta_2^{\text{H}} + 1.965V_x$$
$$[n = 19 \quad p = 0.9696 \quad \text{SD} = 0.160 \quad F = 54.9] \tag{4.15}$$

By combining the two data subsets—that is, combining the phenol, alcohol and steroid data which were collated by Flynn, they produced another model of skin permeability:

$$\log k_p(\text{cm/s}) = -5.048 - 0.586\pi_2^{\text{H}} - 0.633\sum\alpha_2^{\text{H}} - 3.481\sum\beta_2^{\text{H}} + 1.787V_x$$
$$[n = 46 \quad p = 0.9789 \quad \text{SD} = 0.249 \quad F = 235.0] \tag{4.16}$$

Given the statistical quality of this model, they recommend that it is a good generalised model for skin permeability. Further, as the model retains its statistical quality, compared with the models of subsets (Eqs. 4.12–4.15) they suggest that there is no separate mechanism of permeation for each of the same chemical classes and that their permeability can be described by a single equation. They further analysed a series of models based on what they describe as the "log P_{oct} model". In general, they found poorly correlated models, compared to their hydrogen bond-based approach, described above, resulted. Nevertheless, they did find that "better" models (in terms of the correlation coefficient) were obtained when larger data sets were used. They also briefly examined the process of back-diffusion upward through skin layers, suggesting that the regression equation for the ingress of exogenous chemicals (inward rate constant) and log K_m (equilibrium constant) are similar, whereas the rate constant for back-diffusion, log $k_{p\text{back}}$, is very different. This, they suggest, means that log $k_{p\text{back}}$ is wholly or partially diffusion controlled. They base this on the comparative size of coefficients in Eqs. 4.13 and 4.17, which represents back-diffusion, below:

$$\log k_p(\text{cm/s}) = -5.304 - 0.247\pi_2^{\text{H}} - 0.716\sum\alpha_2^{\text{H}} - 1.670\sum\beta_2^{\text{H}} + 0.018V_x$$
$$[n = 22 \quad p = 0.9849 \quad \text{SD} = 0.242 \quad F = 137.5] \tag{4.17}$$

This implies that molecular structural effects have a smaller effect on log $k_{p\text{back}}$ than on the k_p.

Abraham and colleagues followed up this comprehensive study by later considering the nature and issue of outliers within the data sets, by examining issues with potentially erroneous values for some of the steroid data (Abraham et al. 1997, 1999). They stated, in the first of these papers, that although algorithms based on log$P$ (or logK_{ow}) and molecular weight (or volume) appeared to have utility, particularly in the prediction of logk_p, these algorithms [those published by Potts and Guy (1992, 1995) and others, related models using the same, or similar data sets, such as Brown and Rossi (1989), Fiserova-Bergerova et al. (1990), McKone and Howd (1992), Guy and Potts (1993) and Wilschut et al. (1995)] were empirical in nature and yielded little information on the structural features of solutes that influenced water–skin permeability. Thus, they continued their solvation-based approach which focused on the use of hydrogen bond descriptors. They initially (Abraham et al. 1997) focused on steroids and included data from Johnson et al. (1995) in their analysis which appeared to be at odds with earlier published work by Scheuplein et al. (1969). Interestingly, they omitted water from their data analysis as, on partitioning from bulk water to skin, they commented that it is not acting as a solute at all. Therefore, using the data they had previously collated from Flynn (1990), and reduced into a smaller, more chemically specific data set, they obtained the following expression:

$$\log k_p(\text{cm/s}) = -5.241 + 0.437R_2 - 0.410\pi_2^{\text{H}} - 1.631\sum\alpha_2^{\text{H}} - 3.286\sum\beta_2^{\text{H}} + 2.012V_x$$
$$[n = 47 \quad r^2 = 0.9567 \quad \text{SD} = 0.197 \quad F = 181] \tag{4.18}$$

They suggest that this algorithm is statistically the best of those which use hydrogen bonding descriptors to describe solute sets that do not include steroids. They then considered the incorporation of the steroid data into the algorithm and suggested that the Johnson data were a closer fit than the Scheuplein data. Thus, they then added five steroids (excluding hydrocortisone, due to the large spread of reported results) from Johnson et al. (1996) and also dexamethasone but not lignocaine (lidocaine; due to issues with its neutral and charged species) to their data set they obtain their final algorithm:

$$\log k_p(\text{cm/s}) = -5.132 + 0.439R_2 - 0.489\pi_2^{\text{H}} - 1.478\sum\alpha_2^{\text{H}} - 3.442\sum\beta_2^{\text{H}} + 1.941V_x$$
$$[n = 53 \quad r^2 = 0.9577 \quad \text{SD} = 0.213 \quad F = 213] \tag{4.19}$$

Equation 4.19 was their preferred equation, due to the statistical quality of fit and the larger molecular space occupied by the data set compared to previous models. Thus, they also suggested that a single algorithm describing permeability [as suggested by Potts and Guy (1992)] was more appropriate than a series of models describing different parts of the permeation process, such as Flynn's suggestion of an aqueous pore pathway for hydrophilic permeants (Flynn 1990). Nevertheless, they did suggest that, for *very* hydrophilic potential permeants, particularly those whose log*P* values are lower than those of methanol and formic acid, the aqueous pore pathway might be a possible route of absorption. In doing so they recommended that Eq. 4.19 is not used to model the permeation of such molecules.

Thus, these two studies by Abraham et al. (1997, 1999) demonstrate the role of hydrogen bonding in the process of skin permeability, and the positive correlation observed in their various algorithms for molecular size (usually as V_x) reflects on the role of hydrophobicity in the permeation process. Further, they suggest that there are issues with the steroid data available in the literature and which has been incorporated into various models.

The issue of the role of hydrogen bonding was also discussed by other studies, including a series of landmark, in-depth studies by Roberts, Hadgraft and Pugh, among others (Roberts et al. 1995, 1996; Pugh et al. 1996, 2000; Pugh 1999).

Roberts et al. (1995) initially analysed the methods available in the literature to predict the skin permeation of monofunctional solutes from aqueous solutions. They excluded polyfunctional molecules because their solvatochromic parameters are calculated by a summation method, but the authors stated that they had no evidence that additivity of effects was justified (Roberts et al. 1995). It is interesting to compare this with the approach by Abraham et al. (1995) who did use such an approach, which they found was related to the chemicals in the data set chosen for analysis.

Roberts et al. determined the permeability coefficients of a range of such compounds ($n = 24$), which generally included chemical functionalities as alcohols, phenols, acids, ethers and nitro functional groups. They then carried out multiple linear regression analysis against a range of widely used and commonly available physicochemical parameters, including partition coefficients and solvatochromic parameters. They undertook seven analyses which related to the following: (1) the lipophilicity (as the octanol–water partition coefficient) and molecular weight, i.e. following the Potts and Guy (1992) method; (2) molecular group contributions based on the ab initio method reported by Pugh and Hadgraft (1994); (3) hydrogen bond donor ability, based on the method reported by El Tayar et al. (1991); (4) penetrant volume ($V - 1$), hydrogen bond donor (α) and hydrogen bond acceptor (β) abilities; (5) a method based solely on hydrogen bond donor ability; (6) a method to derive a model based solely on the solubility parameter reported by Fedors (1974); and (7) the development of a new 2-phase (the "dual solvent" approach) model describing percutaneous absorption for the data set used in this study. They produced a series of models, based on analysing the same data using the different methods listed above. In total, twenty models were produced with the significant caveat that they were developed using a small data set of simple, monofunctional molecules.

They found that the solvatochromic group contribution and the dual-phase models were more successful than the other methods in predicting skin permeability coefficients. They attributed this to both methods using descriptors that represented certain parameters defined in the solvatochromic approach. They determined that they were able to develop reasonably accurate and simple-to-use structure–permeability relationships that were empirical in nature and offered no mechanistic insight to the process of skin permeability. Thus, Roberts et al. (1995) demonstrated that for a common group of 24 penetrants, the available methods for predicting skin permeability were not the same. They attributed this to the methods used in generating each model; for example, the solvatochromic method is a regression model which considers discrete molecular properties (molecular volume and hydrogen bonding) and the Potts and Guy (1992) approach—by other researchers who have adopted their approach—uses methods derived from those considered by El Tayar et al. (1991) where the regression terms represent complex combinations of fundamental physicochemical properties. Thus, they conclude that the solvatochromic approach offers more than the other methods in terms of yielding significant mechanistic information on the process of skin permeation and that in terms of simple screening of compounds as potential permeants the ab initio approach of Pugh and Hadgraft (1994) is simple to use—in that it can be used with only a knowledge of the molecular structure of a penetrant—and produces reasonable predictions of skin permeability.

Their second study (Roberts et al. 1996) focused on the role of hydrogen bonding on skin permeability. They begin by discussing the observation—by Roberts in 1976—that the skin permeability of phenolic compounds was inversely related to the number of hydrogen bonding groups present on a penetrant (Roberts 1976). This subject had been revisited a few years later by Kamlet et al. (1983) and had resulted in the development of the solvatochromic theory, which has been applied to partitioning and, in the study by Abraham et al. (1995), epidermal permeability.

Thus, Roberts et al. constructed a data set of permeability data taken from five previously published literature sources. Following a range of analyses they concluded that, while solute lipophilicity was a major determinant for partitioning into the *stratum corneum*, the hydrogen bonding ability of a solute was the main determinant of solute diffusing across the *stratum corneum*. Diffusion was described by the following algorithm:

$$\begin{aligned} &\log(D/h) = -2.47 - 0.191191\sigma\mathrm{H} - 0.0853\sigma\mathrm{C}^* \\ &[n = 45 \quad r^2 = 0.709] \end{aligned} \tag{4.20}$$

where

D is the diffusion coefficient

h is the path length for diffusion

σH is the number of H-bonding groups present on the permeant

σC* is the number of carbons not involved in a C=O bond in the penetrant.

Equation 4.20 was used in preference to models featuring molecular weight, as the inclusion of the term σC* improved the quality of the regression. The inclusion of solvatochromic parameters for individual hydrogen bonding groups improved further the statistical quality of the model:

$$\begin{aligned} &\log(D/h) = -1.86 - 0.605\alpha - 2.09\beta \\ &[r^2 = 0.904] \end{aligned} \tag{4.21}$$

They validated their models by producing a series of relationships between lag time and diffusivity. In doing so they determined that diffusion across the *stratum corneum* was related to the number of hydrogen bonding groups present on a solute. Thus, if a potential skin permeant had 0–2 hydrogen bonding groups, it has the most pronounced effect on the rate of diffusion, and the addition of additional such groups exerted lesser and lesser effects on diffusion.

The third part of this series again focused on the role of hydrogen bonding in skin permeability, but expanded the scope of their previous study to explore the role of molecular size on skin permeability (Pugh et al. 1996). They described the permeability coefficient, k_p, which is used widely for characterising skin permeation. k_p itself is a composite parameter, describing partitioning and diffusion phenomena; specifically, the partition of a chemical into the outermost layer of the *stratum corneum* from a topical vehicle or formulation (described by K_{SC} and described, from water, by the expression $\log K_{SC} = -0.024 + 0.59 \cdot \log K_{octanol}$) is followed by the diffusion of the chemical across the *stratum corneum* (described by D, the diffusion coefficient across a path length of thickness h). K_{SC} has been related to the octanol–water partition coefficient and suggests the importance of partitioning in the approach to modelling skin absorption taken by Potts and Guy (1992). Thus, Pugh et al. considered that composite physicochemical descriptors such as the partition coefficient may within them encode a number of molecular properties, including hydrogen bonding. Therefore, to decouple these effects they estimated *D*/*h* separately. They described the *stratum corneum* as exhibiting predominately hydrogen bond donor effects (with the ratio of donor to acceptor groups being approximately 60:40) due to the specific chemical functionalities found within the skin and which will influence the nature of hydrogen bonding and consequently the skin permeability, described by the authors as "retardation effects" (RC). These effects manifest themselves as interactions between the hydrogen bonding groups on penetrant and the *stratum corneum*. They developed models which suggested that diffusion could be calculated based on the hydrogen bond groups and nature of a permeant and that the influence of molecular weight was not as significant as reported previously in other studies, in which the degree of interaction or binding—ostensibly by hydrogen bonding—was underestimated. Compared to their previous work polyfunctional molecules were also considered. Thus, for molecules for which α and β were known the RC was attributed to the following expression:

$$\mathrm{RC} = 0.0024 + 1.36(\alpha - \beta) + 3.18\sum\beta \tag{4.22}$$

For polyfunctional molecules where these parameters were not known estimates were made, producing apparent retardation coefficients, RC*. Consequently, they determined that the maximal diffusion, as D_m/h, that could be attained by small molecules which did not bond to the *stratum corneum* was 0.03 cm/h. This value decreases significantly as hydrogen bonding groups were introduced and fell to a very low minimum value after the addition of four groups. Plots of *D*/*h* against RC resembled an adsorption isotherm, and Pugh et al. suggest that Langmuir's equation could be used to fit the data.

A fourth study in this series was published by Pugh et al. (2000). In this study and in a related book chapter (Pugh 1999), prcomponent analysis and multivariate regression analysis were used to assess the physicochemical predictors of skin permeation. They focused on permeant binding to the *stratum corneum* by developing QSAR models to quantify the permeation process. They developed a generalised model in which molecular weight, hydrogen bonding (as the sum of the modulus of partial charges on constituent atoms, as opposed to the use of linear free energy relationships used by, for example Abraham) and electronic charge were considered of significance and in which the best predictors of permeability were molecular weight and charge. Further, by considering the similarity of the eigenvectors and normalised coefficients from their principal component analysis they concluded that charge and molecular weight were equally important parameters in the prediction of diffusion, which sits slightly at odds with the findings of their previous study (Pugh et al. 1996) where the role of molecular weight in diffusion was lower than expected. These somewhat contradictory conclusions, enhanced somewhat by the comments in a related study (Pugh et al. 1998), suggest both the difficulty in interpreting the data used in each study and highlight the difficulty in quantifying hydrogen bonding effects as a discrete phenomenon and not influenced by other physicochemical descriptors. It also highlights the differences in output from different studies, even where *similar* methods and data sets are used.

A range of other models have been reported, mostly based on analyses of the Flynn data set or subsets of data therein. Hostynek and Magee (1997), while importantly considering in vivo formulation and skin effects in their models, also found a strong relationship with hydrogen bonding:

$$\log k_{p\max} = -0.60\text{VEH} - 0.014\text{MR} - 0.16\text{HBA} - 0.29\text{HBD} - 2.1$$
$$[n = 20 \quad r^2 = 0.80 \quad s = 0.35 \quad F = 14.8] \tag{4.23}$$

where
VEH is an indicator variable for the range of vehicles used (i.e. 1 = acetone; 2 = ethanol)
MR is molar refraction
HBA is the number of hydrogen bond acceptors present on a molecule
HBD is the number of hydrogen bond acceptors present on a molecule.

Dearden et al. (2000) analysed the whole Flynn data set, but used a wider range of physicochemical descriptors (81 in total) compared to the previous studies. Their analysis yielded a highly predictive but mechanistically complex QSAR:

$$\log k_p(\text{cm/h}) = -0.626\Sigma\text{Ca} - 23.8\Sigma(Q+)/\alpha - 0.289\text{SsssCH} - 0.0357\text{SsssOH} - 0.482I_\text{B} + 0.405B_\text{R} + 0.834 \tag{4.24}$$
$$[n = 91 \quad r^2 = 0.832 \quad s = 0.563 \quad F = 69.2]$$

where
ΣCa is the HYBOT_PLUS H-bond acceptor free energy factor
$\Sigma(Q+)/\alpha$ is the HYBOT_PLUS positive charge per unit volume
SsssCH is the electrotopological atom-type index for singly bonded CH
SsssOH is the electrotopological atom-type index for singly bonded OH
I_B s the Balaban index
B_R is the number of rotatable bonds.

A number of other studies have examined more homogeneous data, including congeneric series which may be part of the larger Flynn data set. For example, Le and Lippold (1995) examined the in vivo uptake in humans of four homologous esters of nicotinic acid, demonstrating a relationship between penetrant hydrophobicity and uptake. Kitagawa et al. (1997) reported a QSAR based on the permeation across guinea pig dorsal skin for four chemicals (alkyl para-substituted parabens):

$$\log \text{PC (cm/h)} = 0.72 \log K_\text{ow} - 3.26$$
$$[n = 4 \quad r_2 = 0.90 \quad s, F \text{ not given}] \tag{4.25}$$

Further, Kim et al. (2000) reported a quantitative relationship for skin permeability (for rat skin) of testosterone and three of its ester derivatives. Diez-Sales et al. (1993) reported a parabolic relationship between aniline and six 4-alkyl-substituted derivatives; Dal Pozzo et al. (1991) reported a relationship based on hydrophobicity for the permeation of glycol, polyglycol and alkyl esters of nicotinic acid.

What these and a number of other studies all have in common, despite a range of membranes being used, is the comparatively small size of the data sets used for these studies. While all these studies have merit and significance in their own right—and while they mostly strongly re-enforce the relationship between hydrophobicity and permeability—they have little relevance outside the homologous series being explored and, as they often employ very different methods they could not be used to enhance a larger data set, such as that presented by Flynn.

Morimoto et al. (1992) analysed another small data set ($n = 16$) but one which was more structurally diverse. They reported that permeability across human and rat skin was limited for hydrophilic penetrants, almost being constant, and that an increase in hydrophobicity resulted in a linear increase in permeability. Thus, for human skin, they reported the following relationship:

$$k_p = 1.17 \times 10^{-7} \cdot \log K_{\text{ow}}^{0.751} + 2.73 \times 10^{-8} \quad [n = 16 \quad r^2, s, F \text{ not reported}] \tag{4.26}$$

A similar study was reported by Lee et al. (1994), also with sixteen chemicals in their data set. Their structural diversity was not comparable to that reported by Morimoto et al. (1992), so inferences on the permeability of hydrophilic permeants were not able to be made. They obtained a non-linear, biphasic relationship:

$$\log(\%/h)_{\text{invitro}} = 0.07 \log K_{\text{ow}} - 0.04(\log K_{\text{ow}})^2 + 0.81 \quad [n = 16 \in r^2 = 0.84 \quad s, F \text{ not reported}] \tag{4.19}$$

And, for an in vivo study, they reported the following QSAR:

$$\log(\%/h)_{\text{invivo}} = 0.05 \log K_{\text{ow}} - 0.03(\log K_{\text{ow}})^2 + 0.91 \quad [n = 16 \quad r^2 = 0.67 \quad s, F \text{ not reported}] \tag{4.27}$$

A biphasic relationship for the permeation of eight amino acids was also reported by Lin et al. (1996), in which lipophilicity, as determined as the partition coefficient between polyethylene glycol and Na_2SO_4, was found to be significant. Thus, it is clear that very few comparisons between the studies of Lee and Morimoto could be drawn, highlighting the influence of the input data on the model output.

Van Rooij et al. (1995) reported a QSAR based on the absorption of 10 polyaromatic hydrocarbons (PAHs) in which permeability was significantly related to molecular weight. Roy et al. (1998) expanded upon this study and reported a QSAR analysis of the permeability of 60 PAHs in rat skin. They reported a negative relationship between hydrophobicity and permeability but used chemicals which were, overall very hydrophobic, with the lowest $\log K_{ow}$ of their PAHs being 4.00, which may fit into consideration of the biphasic models reported above (essentially being the downslope of such a model), or the comments by Potts and Guy (1992) and Roberts et al. (1995, 1996) on the relative permeability of the *stratum corneum* and the viable epidermis:

$$\begin{aligned} &\mathrm{PADA} = -14.7 \log K_{ow} - 22.0\mathrm{SHDW6} + 111.9 \\ &[n = 60 \quad r^2 = 0.64 \quad s = 7.7 \quad F = 54] \end{aligned} \tag{4.28}$$

where

PADA is the percentage-applied dose that penetrated in vitro rat skin after 24 h
SHDW6 is the Shadow ^ area descriptor for molecular size.

Gute et al. (1999) expanded on this study and concluded that molecular weight by itself was better able to model the percutaneous absorption of PAHs across rat skin than Roy et al's model based on $\log K_{ow}$ and SHDW6:

$$\begin{aligned} &\mathrm{PADA} = -0.3\mathrm{MW} + 90.6 \\ &[n = 60 \quad r^2 = 0.67 \quad s = 7.4 \quad F = 120] \end{aligned} \tag{4.29}$$

A highly significant study was conducted by Magee (1998), which highlighted the complexity of modelling the Flynn data set—or, by inference, similar data sets in this or related fields. In essence, Magee tested the hypothesis that skin permeation changes with the hydrophobicity of the prospective permeant. This was achieved by placing the Flynn data into three overlapping groups, defined by their octanol–water partition coefficients—the ranges were $\log K_{ow}$ values of −1.38 to 1.96, 1.53–2.97 and 2.50–5.49. A difference in the QSARs thus generated was observed, with a different emphasis placed on the significance of $\log K_{ow}$ and molecular weight in each model. Further, for the most hydrophobic group an increased significance in the role of hydrogen bonding was observed. Magee also analysed individual classes of chemicals within the larger data set, and found—somewhat predictably—that correlations for individual chemical classes, which in many cases had been measured in the same laboratory, were better than for the complete data set. While Magee's results shed little new light on the mechanistic underpinnings of skin permeability, they do highlight important issues both in how the available data are analysed and how any such analysis is interpreted (Moss et al. 2002).

Thus, the size of the data set can influence the quality and applicability of a model, the former by the impact the small number of data points has on the statistical quality, particularly variance, of the model, as well as the potential reduction in chemical diversity within the test set, and the latter by the relevance it has particularly in the context of the "molecular space" of the model and how widely it could be applied to estimating the permeability of penetrants which have not already been established. There also appears to be a reasonable comparison between the correlation coefficient (presented as p, r or r^2) which seems to be higher for smaller data sets. Whether this is an artefact of the data set or a feature of the improved, possibly streamlined, nature of the data examined is unclear. Indeed, the nature of the subset itself is of interest, as they are chosen to reflect a possible reduction in experimental error by taking a subset from a single study or single laboratory, or they try to adhere to particular chemical features, such as structural functions or pharmacological classes. For example, Lien and Gao (1995) used a subset of Flynn's data set ($n = 22$) and Potts and Guy's 1995 study used a subset from the same primary source of 37 non-electrolytes. Barratt (1995) examined a subset—comparatively large compared to most other such studies—of Flynn's data which consisted of 60 "small molecules and steroids". In developing a large number of very-well-correlated models, Abraham and co-workers developed models with data sets ranging from, for example, 19–53 compounds (Abraham et al. 1995, 1997, 1999).

Coupled to the size of the available—or chosen—data set is the issue of outliers, particularly their consistent identification and removal. The reason for removing particular outliers is often not clearly justified in the majority of studies and, whether by an arbitrary statistical test or subjective assessment of the data, such model refinement should be transparent and, where possible, consistent. A number of studies, including some of those described above, do indeed justify their decision to remove outliers (Pugh and Hadgraft 1994; Barratt 1995; Cronin et al. 1999) and, in the majority of cases, some chemicals are found consistently to be outliers. The removal of outliers should also be considered in the context of the findings presented by Johnson et al. (1995) and Degim et al. (1998). Johnson et al. (1995) re-examined the permeability coefficients and reported large discrepancies between the steroid permeability data from Scheuplein et al. (1969) that were included in the Flynn (1990) data set. This was summarised by Moss et al. (2002) and is shown in Table 4.1. These data show, for example, an order-of-magnitude difference between the permeability value of estradiol reported by Scheuplein et al. (1969) and collated by Flynn (1990), and eight other values collated by Johnson et al. (1995).

Degim et al. remeasured some permeability data, including three compounds found previously by other researchers to be outliers—naproxen, atropine and nicotine. As with Johnson's review of the Scheuplein/Flynn steroid data, Degim and colleagues found these data to be substantially different. Thus, the removal of such data (by, for example, Cronin et al. 1999) is clearly justified. However, it does pose questions to the validity of the assumptions made with this potentially erroneous data in a large number of studies, some of which use very small data sets, or subsets, where the potential for error is clearly magnified. Moss et al. (2002) therefore recommended that future work with the Flynn data set should include the new data proposed by Johnson et al. (1995) or remove it from any analysis—at the very least, the original data should be treated with caution as they could exert a detrimental effect on the validity of any resultant model.

Thus, it often appears as if removing some data from the model produces a better model. While it is not inherently wrong to do this, one must therefore question the

Table 4.1 Steroid permeabilities through skin (modified from Johnson et al. 1995)

Steroid	Molecular weight	K_{ow}^{a}	Permeability coefficients
Aldosterone	360.4	12	3.0×10^{-6b}
			5.8×10^{-5c}
Hydrocortisone	362.5	34	3.0×10^{-6c}
			1.6×10^{-5b}
			1.2×10^{-4c}
			2.3×10^{-4c}
Cortisone	360.5	26	1.0×10^{-5b}
Estriol	288.4	295	4.0×10^{-5c}
Corticosterone	346.5	87	6.0×10^{-5b}
			3.0×10^{-4c}
Cortexolone	346.5	331	7.5×10^{-5b}
Estradiol	272.4	7200	3.0×10^{-4b}
			3.2×10^{-3c}
			3.4×10^{-3c}
			3.5×10^{-3c}
			4.0×10^{-3c}
			4.1×10^{-3c}
			5.2×10^{-3c}
			5.4×10^{-3c}
			6.1×10^{-3c}
Hydroxyprogesterone	330.5	550	6.0×10^{-4b}
Testosterone	288.4	2042	4.0×10^{-4b}
			2.2×10^{-3c}
Cortexone	330.5	759	4.5×10^{-4b}
Hydroxypregnenolone	330.5	1000	6.0×10^{-4b}
Pregnenolone	316.5	1349	1.5×10^{-3b}
Progesterone	314.5	5888	1.5×10^{-3b}
			1.3×10^{-2c}
			3.0×10^{-2c}
Estrone	270.4	575	3.6×10^{-3c}

Note [a]The octanol–water partition coefficient; [b]The permeability coefficients determined by Scheuplein et al. (1969); [c]The permeability coefficients determined by other researchers, as collated by Johnson et al. (1995)

applicability of the model both in terms of offering a broad—or even single—insight to the mechanism of skin permeation and the predictive power of any model. As the examples above show, this is of particular importance when considering small data sets (i.e. data sets containing approximately 50 or fewer members). However, there is always a careful balance to be struck between improving the quality of the model and reducing its range of applicability and, in such a context the removal of chemicals considered—statistically or otherwise—to be outliers to a good model is subjective and should be undertaken with caution and transparency.

Moss and Cronin (2002) collated a revised version of Flynn's data set which took into account issues raised by the above studies. They derived a QSAR for a data set of 119 compounds—three of which (sucrose, etorphine and digitoxin) were underpredicted by the initial model and removed as outliers based on their residuals. This resulted in the following highly significant expression:

$$\log k_p(\text{cm/s}) = 0.74 \log P - 0.0091\text{MW} - 2.39$$
$$\Big[n = 116 \quad r^2_{(\text{adj})} = 0.82 \quad r^2_{(\text{CV})} = 0.81 \quad F = 266 \tag{4.30}$$
$$\text{t-values: } \log P\,22.8;\ \text{MW} -16.6;\ P < 0.0001 \text{ for both variables}\Big].$$

While this model is similar to those published previously, and discussed above, the statistical analysis carried out indicates that steroids are no longer outliers in this model and should not be considered as such in terms of their skin permeation as they appear to permeate the skin by the same means as the majority of exogenous chemicals in this model. This study was expanded upon by Patel et al. (2002) who developed QSARs using a data set of 158 members. Following the removal of outliers, the following relationship was established:

$$\log k_p = 0.652 \log K_{\text{ow}} - 0.00603\text{MW} - 0.623\text{ABSQon} - 0.313\text{SsssCH} - 2.30$$
$$[n = 143 \quad r^2 = 0.90 \quad s = 0.35 \quad F = 312$$
$$\text{t-values: } \log K_{\text{ow}}\,17.8;\ \text{MW } 7.04;\ \text{ABSQon } -4.48;\ \text{SsssCH } -12.4] \tag{4.31}$$

where
ABSQon is the sum of absolute charges on oxygen and nitrogen atoms
SsssCH is the sum of E-state indices for all methyl groups.

The results of this study confirm that they key descriptors which influence the skin permeability of exogenous chemicals are penetrant lipophilicity, molecular size and hydrogen bonding ability, and confirm the findings of a number of studies reported above. The authors also suggest that molecular volume is the most efficient parameter for characterising the steric properties of molecules thus modelled.

One issue often not considered or reported in the above models is the ionisation state of the penetrant in the aqueous donor solution, which may be unclear and impact the permeability coefficient (k_p) calculated from the flux data. Thus, as Moss et al. (2002) commented, the design of an experiment can influence significantly the output of a QSAR-type study. This may relate, for example, to the potential for ionisation to occur in an experiment, which may mean that more than one species of a chemical can be present. This would usually result in species with potentially different permeabilities being present in the experiment and that may affect the saturated solubility required in the donor phase and also the final permeability coefficient determined as a result of the experiment.

When such "input factors"—including ionisation and issues regarding the quality of data or inconsistencies in published data—are combined with the inherent and expected biological variation associated with such experiments, Mitragotri et al. (2011) commented that the use of more complicated molecular structure descriptors often cannot be justified from a statistical point of view. They did not expand upon this comment, nor furnish examples of where this might be an issue, although Geinoz et al. (2004) did comment that such models with a wide range of very specific parameters might lead to misinterpretation, providing unclear mechanistic insight. However, the main issue in using such parameters lies not in an ambiguous interpretation of their statistical quality but of their direct and simple application in predicting permeability by those researching percutaneous absorption, as such parameters are often calculated with costly and specialist software. Far wider issues, addressed above in the context of literature data (Scheuplein et al. 1969; Flynn 1990; Johnson et al. 1995; Degim et al. 1998; Moss and Cronin 2002), are that removing the ambiguous literature data will have a far more substantial effect on improving the quality of the models and their mechanistic meaning and, secondly, the reduction in size of data sets reduces model applicability and, possibly, statistical relevance. It is therefore relatively easy to misinterpret the findings of Bouwman et al. (2008). They determined that only four published models met their defined criteria for acceptability—those published by ten Berge (2011), McKone and Howd (1992), Moss and Cronin (2002), and Magnusson et al. (2004). These four models all use simple representations of permeability, largely being described by lipophilicity and molecular size, but this does not preclude the relevance of approaches which use alternative methods and reflects positively on the nature of how the study was conducted, rather than a justification of the use of certain descriptors for analysis.

With regard to Chap. 2, and the derivation of experimental data, it is rare that a model of skin absorption examines flux as most models focus on the use of k_p as it is potentially better at comparing chemicals with different properties, providing a standardisation of what might be a wide range of aqueous solubilities. However, flux, often represented as J, $J_{\max}$ or $J_{\text{ss(steady-state)}}$, is more relevant as it describes the amount of chemical absorbed, per unit area, over a defined time period (i.e. it defines the dose of a topically applied pharmaceutical product, such as a transdermal patch, with units of concentration/surface area/time). In terms of building a model that uses J rather than k_p, the former can be estimated from the latter and experimental values of the aqueous solubility of penetrants. Secondly, it may be determined from the following expression:

$$J_{\max} = k_p \cdot C_{v,\text{sat}} \tag{4.32}$$

where $C_{v,\text{sat}}$ is the saturated solubility of the penetrant in its vehicle; $C_{v,\text{sat}}$ and k_p are normally determined in the same vehicle.

Magnusson et al. (2004) developed a QSAR model for $J_{\max}$ rather than k_p. They found that $\log K_{\text{ow}}$ was less significant and that molecular weight alone described the bulk a penetrant-dependent variance in the data. Further studies by this group (i.e. Zhang et al. 2009) extended this approach and will be discussed in Chaps. 6 and 7.

The years following the publication of key works by Flynn (1990) and Potts and Guy (1992) saw a plethora of related studies published. They all use the Flynn data set—or subsets of Flynn's data set—or related data sets to describe the process of skin permeation and, in doing so via descriptive algorithms featuring discrete parameters, provided simple (in some cases) predictive models that allowed the estimation of skin permeability for new chemicals. The models also described the process of skin permeation with varying degrees of mechanistic insight, although the extent of this depended on the nature of the descriptors used in the initial modelling and the composition of the data set being analysed.

Nevertheless, the models developed to this stage have, to different degrees, had significant limitations. The larger, "general" models are compiled from a range of sources as the data have been generated in a number of different laboratories. While this may generate variability in the input data which is reflected in a poor output, a number of researchers have commented that such interlaboratory variation is expected in skin permeability experiments and is difficult to decouple from the inherent variation associated with the skin barrier and with skin permeability. Such variance is carried into models thus developed and may manifest itself in the statistical quality of the model.

Perhaps the most obvious limitation relates to formulation. Although Flynn described skin permeability as occurring at the skin interface from aqueous solutions, this does not describe vehicle effects, either on partitioning of potential permeants from them or the effect that they may exert on the integrity of the skin barrier. Attempts to describe skin permeation in such circumstances will be discussed in later chapters.

A further limitation of the models is the endpoint used in describing skin permeability. Magnusson et al. (2004) is one of the few studies that use anything other than a variation of the permeability coefficient (k_p) in describing skin absorption. However, the development of models where endpoints other than dermal uptake—however, it is notated—is limited. Thus, models may estimate skin permeability, but they current do not estimate skin deposition, which may be more important than permeability when application to the skin of local therapies is considered. This may extend into models which focus on biological endpoints within the skin, such as those which may characterise and potentially quantify processes such as skin irritation. Thus, the potential for expanding the use of the methods described in this chapter is still enormous, particularly in moving beyond the current in vitro limits of most models.

If the above description of skin permeability studies and their resultant QSAR-based analyses indicates anything, it is that new skin permeability data are required, both in terms of quality/consistency and quantity. This issue will be discussed throughout later chapters. However, the time-consuming nature of generating new data, which in many cases, might be unsuitable for inclusion in a model as it is focused on the needs of a particular study rather than the wider and possibly different needs of a model and is a significant limitation to the development of this field. So too is the standardisation in the use of methods to analyse such data. Finally, it is clear that the development of models that estimate skin permeability is hindered by their inappropriate use—for example, the use of a model outside the range of the data set from which it was constructed, or the application of a model generated for a small, specific data set to inappropriate data.

References

Abraham MH (1993) Scales of hydrogen bonding: their construction and application to physicochemical and biochemical processes. Chem Soc Rev 22:73–83

Abraham MH, Whiting GS, Doherty RM, Shuely WJ (1990) Hydrogen bonding part 13. A new method for characterisation of GLC stationary phases—the Laffort effect. J Chem Soc Perkin Trans 2:1451–1460

Abraham MH, Whiting GS, Doherty RM, Shuely WJ (1991) Hydrogen bonding part 16. A new solute solvation parameter pi-h2 from gas chromatographic data. J Chromatogr 587:213–228

Abraham MH, Chadha HS, Mitchell RC (1995) The factors that influence skin penetration of solutes. J Pharm Pharmacol 47:8–16

Abraham MH, Martins F, Mitchell RC (1997) Algorithms for skin permeability using hydrogen bond descriptors: the problem of steroids. J Pharm Pharmacol 49:858–865

Abraham MH, Chadha HS, Martins F, Mitchell RC, Bradbury MW, Gratton JA (1999) Hydrogen bonding part 46. A review of the correlation and prediction of transport properties by an LFER method: physicochemical properties, brain penetration and skin permeability. Pest Sci 55:78–88

Anderson BD, Higuchi WI, Raykar PV (1988) Heterogeneity effects on permeability-partition coefficient relationships in human stratum corneum. Pharm Res 5:566–573

Barratt MD (1995) Quantitative structure-activity relationships for skin permeability. Toxicol In Vitro 9:27–37

Bouwman T, Cronin M, Bessems J, van de Sandt J (2008) Improving the applicability of (Q)SARs for percutaneous penetration in regulatory risk assessment. Hum Exp Toxicol 27:269–276

Brown SL, Rossi JE (1989) A simple method for estimating dermal absorption of chemicals in water. Chemosphere 19:1989–2001

Bunge AL, Cleek RL (1995) A new method for estimating dermal absorption from chemical exposure. 2. Effect of molecular weight and octanol-water partitioning. Pharm Res 12:88–95

Bunge AL, Cleek RL, Vecchia BE (1995) A new method for estimating dermal absorption from chemical exposure. 3. Compared with steady-state methods for prediction and data analysis. Pharm Res 12:972–982

Carslaw HS, Jaeger JC (1980) Conduction of heat in solids, 2nd edn. Clarendon Press, Oxford

Cleek RL, Bunge AL (1993) A new method for estimating dermal absorption from chemical exposure. 1. General approach. Pharm Res 10:497–506

Crank J (1975) The mathematics of diffusion, 2nd edn. Clarendon Press, Oxford

Cronin MTD, Dearden JC, Moss GP, Murray-Dickson G (1999) Investigation of the mechanism of flux across human skin in vitro by quantitative structure-permeability relationships. Eur J Pharm Sci 7:325–330

Dal Pozzo AD, Donzelli G, Liggeri E, Rodriguez L (1991) Percutaneous absorption of nicotinic acid derivatives in vitro. J Pharm Sci 80:54–57

Dearden JC, Cronin MTD. Patel H., Raevsky OA (2000) QSAR prediction of human skin permeability coefficients. J Pharm Pharmacol 52S:221

Degim IT, Pugh WJ, Hadgraft J (1998) Skin permeability data: anomalous results. Int J Pharm 170:129–133

Diez-Sales O, Perez-Sayas E, Martin-Villodre A, Herraez-Dominguez M (1993) The prediction of percutaneous absorption. I. Influence of the dermis on in vitro permeation models. Int J Pharm 100:1–7

El Tayar N, Tsai RS, Testa B, Carrupt PA, Hansch C, Leo A (1991) Percutaneous penetration of drugs—a quantitative structure-permeability relationship study. J Pharm Sci 80:744–749

Fedors RF (1974) A method for estimating both the solubility parameters and molar volumes of liquids. Polymer Eng Sci 14:147–154

Fiserova-Bergerova V, Pierce JT, Droz PO (1990) Dermal absorption potential of industrial chemicals: criteria for skin notation. Am J Ind Med 17:617–635

Flynn GL (1990) Physicochemical determinants of skin absorption. In: Gerrity TR, Henry CJ (eds) Principles of route-to-route extrapolation for risk assessment. Elsevier, New York, pp 93–127

Geinoz S, Guy R, Testa B, Carrupt P (2004) Quantitative structure-permeation relationships (QSPeRs) to predict skin permeation: a critical evaluation. Pharm Res 21:83–92

Gute BD, Grunwald GD, Basak SC (1999) Prediction of the dermal penetration of polycyclic aromatic hydrocarbons (PAHs): a hierarchical QSAR approach. SAR QSAR Environ Res 10:1–15

Guy RH, Potts RO (1993) Penetration of industrial chemicals across the skin: a predictive model. Am J Ind Med 23:711–719

Hostynek JJ, Magee PS (1997) Modelling in vivo human skin absorption. Quant Struct-Act Relat 16:473–479. J. Pharm. Sci. 1994: 4, 562–565

Johnson ME, Blankstein D, Langer R (1995) Permeation of steroids through human skin. J Pharm Sci 84:1144–1146

Johnson ME, Mitragotri S, Patel A, Blankstein D, Langer R (1996) Synergistic effects of chemical enhancers and therapeutic ultrasound on transdermal drug delivery. J Pharm Sci 85:670–679

Kamlet MJ, Abboud JL, Abraham MH, Taft RW (1983) Linear solvation energy relationships. 23. A comprehensive collection of the solvatochromic parameters, π^*, α, and β, and some methods for simplifying the generalized solvatochromic equation. J Organ Chem 48:2877–2887

Kim MK, Lee CH, Kim DD (2000) Skin permeation of testosterone and its ester derivatives in rats. J Pharm Pharmacol 52:369–375

Kitagawa S, Li H, Sato S (1997) Skin permeation of parabens in excised guinea pig dorsal skin, its modification by penetration enhancers and their relationship with n-octanol/water partition coefficients. Chem Pharm Bull 45:1354–1357

Le VH, Lippold BC (1995) Influence of physicochemical properties of homologous esters of nicotinic acid on skin permeability and maximum flux. Int J Pharm 124:285–292

Lee CK, Uchida T, Kitawga K, Yagi A, Kim NS, Goto S (1994) Skin permeability of various drugs with different lipophilicity. J Pharm Sci 4:562–565

Lien EJ, Gao H (1995) QSAR analysis of skin permeability of various drugs in man as compared to in vivo and in vitro studies in rodents. Pharm Res 4:583–587

Lin RY, Hsu CW, Chen WY (1996) A method to predict the transdermal permeability of amino acids and dipeptides through porcine skin. J Controlled Release 38:229–234

Magee PS (1998) Some novel approaches to modelling transdermal penetration and reactivity with epidermal proteins. In: Devillers J (ed) Comparative QSAR. Taylor & Francis, London, pp 137–168

Magnusson BM, Anissimov YG, Cross SE, Roberts MS (2004) Molecular size as the main determinant of solute maximum flux across the skin. J Invest Dermatol 122:993–999

McKone TE, Howd RA (1992) Estimating dermal uptake of nonionic organic chemicals from water and soil: I. Unified fugacity-based models for risk assessments. Risk Anal 12:543–557

Mitragotri S, Anissimov YG, Bunge AL, Frasch HF, Guy RH, Hadgraft J, Kasting GB, Lane ME, Roberts MS (2011) Mathematical models of skin permeability: an overview. Int J Pharm 418:115–129

Morimoto Y, Hatanaka T, Sugibayashi K, Omiya H (1992) Prediction of skin permeability of drugs: comparison of human and hairless rat skin. J Pharm Pharmacol 44:634–639

Moss GP, Cronin MTD (2002) Quantitative structure-permeability relationships for percutaneous absorption: re-analysis of steroid data. Int J Pharm 238:105–109

Moss GP, Dearden JC, Patel H, Cronin MTD (2002) Quantitative structure-permeability relationships (QSPRs) for percutaneous absorption. Toxicol In Vitro 16:299–317

Ozisik MN (1980) Heat conduction. Wiley, New York

Patel H, ten Berge W, Cronin MTD (2002) Quantitative structure-activity relationships (QSARs) for prediction of skin permeation of exogenous chemicals. Chemosphere 48:603–613

Potts RO, Guy RH (1992) Predicting skin permeability. Pharm Res 9:663–669

Potts RO, Guy RH (1995) A predictive algorithm for skin permeability: the effects of molecular size and hydrogen bond activity. Pharm Res 12:1628–1633

Pugh WJ (1999) Relationship between H-bonding of penetrants to stratum corneum lipids and diffusion. In: Bronaugh RL, Maibach HI (eds) Percutaneous absorption: Drugs–Cosmetics–Methodology. Dekker, New York, pp 177–192

Pugh WJ, Hadgraft J (1994) Ab initio prediction of human skin permeability coefficient. Int J Pharm 103:163–178

Pugh WJ, Roberts MS, Hadgraft J (1996) Epidermal permeability—penetrant structure relationships. 3. The effect of hydrogen bonding interactions and molecular size on diffusion across the stratum corneum. Int J Pharm 138:149–165

Pugh WJ, Hadgraft J, Roberts MS (1998) Physicochemical determinants of stratum corneum permeation. In: Roberts MS, Walters KA (eds) Dermal absorption and toxicity assessment. Drugs in the pharmaceutical sciences, vol 91. Marcel Dekker, London, pp 245–268

Pugh WJ, Degim IT, Hadgraft J (2000) Epidermal permeability—penetrant structure relationships. 4. QSAR of permeant diffusion across human stratum corneum in terms of molecular weight, H-bonding and electronic charge. Int J Pharm 197:203–211

Raykar PV, Fung MC, Anderson BD (1988) The role of protein and lipid domains in the uptake of solutes by stratum corneum. Pharm Res 5:140–150

Roberts MS (1976) Percutaneous absorption of phenolic compounds. Ph.D. thesis, University of Sydney, Australia

Roberts MS, Anderson RA, Swarbrick J (1977) Permeability of human epidermis to phenolic compounds. J Pharm Pharmacol 29:677–683

Roberts MS, Pugh WJ, Hadgraft J, Watkinson A (1995) Epidermal permeability—penetrant structure relationships. 1. An analysis of methods of predicting penetration of monofunctional solutes from aqueous solutions. Int J Pharm 126:219–233

Roberts MS, Pugh WJ, Hadgraft J (1996) Epidermal permeability—penetrant structure relationships. 2. The effect of H-bonding groups in penetrants on their diffusion through the stratum corneum. Int J Pharm 132:23–32

Roy TA, Krueger AJ, Mackerer CR, Neil W, Arroyo AM, Yang JJ (1998) SAR models for estimating the percutaneous absorption of polynuclear aromatic hydrocarbons. SAR QSAR Environ Res 9:171–185

Scheuplein RJ, Blank IH (1971) Permeability of the skin. Physiol Rev 51:702–747

Scheuplein RJ, Blank IH, Brauner GI, MacFarlane DJ (1969) Percutaneous absorption of steroids. J Invest Dermatol 52:63–70

ten Berge W (2011) QSARs for skin permeation of chemicals. Available at: http://home.planet.nl/~wtberge/qsarperm.html. Accessed 5 Aug 2014

van Rooij JGM, Vinke E, de Lange J, Bruijnzeel PLB, Bodelier-Bade MM, Noordhoek J, Jongeneelen FJ (1995) Dermal absorption of polycyclic aromatic hydrocarbons in the blood-perfused pig ear. J App Toxicol 15:193–200

Wilschut A, ten Berge WF, Robinson PJ, McKone TE (1995) Estimating skin permeation: the validation of five mathematical models. Chemosphere 30:1275–1296

Zhang Q, Grice JE, Li P, Jepps OG, Wang GJ, Roberts MS (2009) Skin solubility determines maximum transepidermal flux for similar size molecules. Pharm Res 26:1974–1985

Chapter 5
Algorithms for Estimating Permeability Across Artificial Membranes

The Role of Artificial Membranes in Studies of Percutaneous Absorption

As discussed in Chap. 2, there are a range of established and validated in vitro methods for the measurement of percutaneous absorption. In general, in vitro experiments of the nature described in Chap. 2 will form a significant part of early-stage evaluation of pharmaceutical formulations or in risk assessment protocols. Their use is followed by, and informs, preclinical and clinical evaluation. While fresh human skin (either as full thickness skin, heat-separated epidermal tissue or skin dermatomed to a defined thickness) is the perceived "gold standard" for in vitro testing, it is not always available and certain well-defined compromises are commonly adopted, including the use of human skin that had previously been frozen. Moving further "backwards" from the idealised in vitro model leads to the use of animal tissue; while the use of tissue from a range of species (rat, mouse, pig, guinea pigs, snakes and various species of monkey) has been widely reported in the literature, it is accepted that pigskin is the best model for human skin, with the pig ear being widely used despite differences in the lateral packing of *stratum corneum* lipids and suggestions that it may have a lower barrier function than human skin (Petitot et al. 2007; Vallet et al. 2007; Caussin et al. 2008; Klang et al. 2012). In order to address the issue of tissue variation and availability, various cultured skin alternatives, based on the living skin equivalent models, have also been considered. This technology includes marketed products such as EpiDerm®, EpiSkin® and SkinEthic®. Reconstructed skin models have also been considered although they have been found to exhibit higher permeability than excised mammalian skin as they often have an incomplete or inconsistent barrier (Van Gele et al. 2011; Kuchler et al. 2013). In general, their use has not become widespread, and they have a peripheral role in the models of skin absorption (Netzlaff et al. 2005; Schafer-Korting et al. 2008).

G.P. Moss et al., *Predictive Methods in Percutaneous Absorption*,
DOI 10.1007/978-3-662-47371-9_5

Thus, despite the scientific limitations and logistical constraints discussed above, artificial membranes have found widespread use in early-stage assessment of percutaneous absorption. It is not the aim of this chapter to review these studies, but a few examples are given below, and present an important context for consideration of model development. For example, Ahmed et al. (1983) characterised phenothiazine transport across liquid–lipid, phospholipid and soft polymer membranes. Feldstein et al. (1998) carried out a comparative study of human skin permeability and permeability across a "skin-imitating" PDMS–polycarbonate block copolymer (Carbosil®). They used a group of 14 drugs with diverse therapeutic and physicochemical properties. They found that their two-phase artificial membrane exhibited similar diffusion characteristics as human skin for their 14 penetrants. In a similar study, Shumilov et al. (2009) also evaluated a biphasic artificial membrane. However, neither membrane has found widespread use.

Woolfson et al. (1998) examined a range of tetracaine formulations and investigated their permeation across a PDMS (Silastic®) membrane. They commented that, in cases where the lipophilicity of the penetrant was the prime determinant of drug flux, which is the case for the lipophilic local anaesthetic tetracaine (amethocaine), PDMS membranes had been shown to produce good correlations with the in vivo situation and had proven particularly useful in the development of local anaesthetic systems (Woolfson et al. 1988; Woolfson and McCafferty 1993). Woolfson's 1998 study also correlated reasonably well with a later study using porcine skin (Moss et al. 2006). Other studies, for example Khan et al. (2005) and Kumprakob et al. (2005), also used silicone membranes to assess drug delivery, with the former study comparing permeability across a silicone membrane to pigskin permeability and observing significant differences in the distribution of the permeability across both membranes. Wasdo et al. (2009) also found correlations between PDMS and mammalian skin permeability, developing a series of models to quantify their findings for a 32-member data set. Similarly, Gullick et al. (2010) found reasonable correlations between in vitro diffusion experiments using PDMS membranes and pigskin.

Further, several researchers have used artificial membranes, mostly polydimethylsiloxane (PDMS), to investigate the mechanisms of membrane transport (Waktinson et al. 1994; Pellett et al. 1994). Ley and Bunge (2007) used PDMS membranes to compare permeation from finely divided pure powder and saturated aqueous solutions of model penetrants and examining the role of surface coverage in particular. Dias et al. (2007) used PDMS membranes to compare the release characteristics of saturated solutions due to their homogeneity and uniformity, compared to mammalian skin. They found that permeability was related to the physicochemical properties of their penetrants (i.e. the comparative log *P* values of caffeine, salicylic acid and benzoic acid were reflected in their permeation rates) and that the solvents were taken up into the membrane, altering its properties and the flux of the permeants. They concluded that membrane flux is governed by a combination of solvent and solute characteristics, including size, shape and charge distribution. ATR-FTIR spectroscopy was used to evaluate diffusion across a PDMS membrane (McAuley et al. 2009). Diffusion was described by a Fickian model, and it was determined that the three model chemicals examined—cyanophenol, methyl nicotinate and butyl paraben—all diffused across the membrane independently from the solvent. In one case, a solvent–solute bonded complex of cyanophenol and isostearyl isostearate was observed. The relative diffusion rates of the different permeants were generally attributed to molecular size. McAuley et al. (2010) also developed a rudimentary structure–activity relationship for permeability across a PDMS membrane. Olivera et al. (2010) also used a thermodynamic and kinetic analysis of temperature-dependent PDMS diffusion to elucidate the possible mechanisms of transport. They found a break point for butanol which appeared to differentiate mechanisms of solute diffusion and partitioning which was potentially associated with temperature-induced changes in the properties of the solvent, underlining the significance of temperature control in such experiments.

However, Moss et al. (2006) examined a wide (in terms of their physicochemical properties) range of prodrugs of captopril, characterising their permeability across pigskin and a PDMS membrane. They found a biphasic relationship between molecular properties (notably log *P* and MW) where skin permeability increased with increases in log *P* and MW and then decreased for larger, lipophilic molecules. In significant contrast, permeability across the Silastic® membrane increased exponentially as log *P* and MW were increased. Poor correlations were therefore found between the Silastic® membrane and pigskin permeability. This sits somewhat at odds with a number of other studies, some of which are described above, and is primarily due to the wide range of physicochemical properties examined by Moss et al., compared to the majority of other studies which used narrower molecular spaces in making their comparisons. In most cases, comparisons were made for membrane permeability for one chemical or a series of similar chemicals, such as drugs in a similar therapeutic class.

Frum et al. (2007) used five model penetrants to examine the normal distribution of permeability coefficients across a PDMS membrane. Their findings—that the permeability coefficients of all five drugs were distributed in a Gaussian-normal fashion—are in contrast with those reported for mammalian skin, which were found to be non-Gaussian in a number of studies reviewed by Frum et al. (Liu et al. 1991; Williams et al. 1992; Cornwell and Barry 1995; Kasting et al. 1992; Watkinson et al. 1998; Roper et al. 2000; Fasano et al. 2002; Khan et al. 2005; Wenkers and Lippold 1999), in which log-normal patterns were common. They attributed this difference to the heterogeneity of biological membranes, including the possibility of multiple permeation pathways in mammalian skin, which is in stark contrast to the homogeneity of PDMS, and similar, membranes.

Therefore, while significant limitations have been identified in the use of such membranes (i.e. Moss et al. 2006), artificial membranes can provide an effective screen in early-stage formulation development, and given the lack of biological variation, valuable mechanistic information can be obtained from permeation studies employing such membranes. Therefore, there is significant value in developing quantitative models which describe permeability across such membranes, particularly in comparing them to models of mammalian skin transport.

Quantitative Models for Permeability Across Polydimethylsiloxane Membranes

Given the early contribution of Potts and Guy (1992) in providing a robust quantitative model for human skin permeability, it is perhaps not surprising that work on similar models for membranes other than human skin has lagged behind somewhat. The first major studies quantifying permeability across a PDMS membrane were reported by Chen et al. (1993, 1996). In their first study, they developed empirical models for permeation across a PDMS membrane for 103 chemicals which related flux through the PDMS membrane to partial atomic charge, mole fraction solubility and molecular weight:

$$\begin{aligned}\log J_{\mathrm{mss}} = {} & 0.256 - 4.176\sum e_H - 1.388\sum e_p + 3.807\left(\sum e_H \cdot \sum e_p\right) \\ & + 0.634\log \mathrm{MF} - 0.008\,\mathrm{MW} - 0.753\,\mathrm{imidazole} + 0.626\,\mathrm{amine} \\ & \left[n = 103 \quad r^2 = 0.972 \quad s = 0.217 \quad F = 468.3\right]\end{aligned} \tag{5.1}$$

where

J_{mss} is the maximum steady-state flux (μ mol/s/cm^2);
J_{mss} is the maximum steady-state flux (μ mol/s/cm^2);
e_H is the charge value on a hydrogen with charge higher than 0.1;
e_p is the absolute charge value of a heteroatom which contains unshared electron pairs in the outer shell and all of which are unconjugated;
MF is the mole fraction solubility of a diffusant in isopropyl alcohol;
MW is the molecular weight (g/mol); and

Imidazole and amine are indicator variables for the imidazole and aliphatic amine groups.

Consideration of Chen's initial QSPR in the context of maximum flux shows that the mole fraction term in Eq. 5.1 is related to the solubility (C_s) term in this expression and all other terms are related to membrane permeability. They commented that the partition coefficient and the diffusion coefficient both depend on the solute–solvent–membrane interaction, a finding in common with the findings of Hadgraft and colleagues, discussed above.

In their second such study, Chen et al. (1996) examined a larger data set and refined Eq. 5.1:

$$\begin{aligned}\log J_{\mathrm{mss}} = {} & -2.497 - 4.339\sum e_+ - 1.531\sum e_- + 4.065\left(\sum e_+ \cdot \sum e_{p-}\right) \\ & + 0.649\log C_S - 0.00651\,\mathrm{MW} - 0.640\,\mathrm{imidazole} + 0.689\,\mathrm{amine} \\ & \left[n = 103 \quad r^2 = 0.966 \quad s = 0.238 \quad F = 386.5\right]\end{aligned} \tag{5.2}$$

where

J_{max} is the maximum steady-state flux (μ mol/s/cm^2);

Σe_+ is the sum of the charge values of hydrogen atoms with charge higher than 0.1 and the positive charge of a nitrogen atom in a nitro group; and

Σe_- is the sum of the absolute charge values of all other heteroatoms with unshared electron pairs in the same molecule.

Chen et al. reported that Eq. 5.2 gave better predictions than their previous model (Eq. 5.1; Chen et al. 1993). Thus, they applied Eq. 5.2 to predict the flux of 171 new compounds which were not included in their previous study. This analysis yielded a simplified model in which the imidazole descriptor is not included:

$$\log J_{\text{mss}} = -2.497 - 4.339\sum e_+ - 1.531\sum e_- + 4.065\left(\sum e_+ \cdot \sum e_-\right) + 0.649 \log C_s - 0.00651\,\text{MW} + 0.689\,\text{amine} \tag{5.3}$$

While Chen's studies examined in detail the various subclasses in their data sets, they did not apply this analysis to the whole data set. Although the models are statistically highly relevant, they require the measurement of specific properties, such as the solubility of permeants in isopropyl alcohol as a method does not currently exist to compute this value. Therefore, Cronin et al. (1998) reanalysed the data published by Chen, with the aim of developing QSAR models based on readily calculable descriptors and with greater mechanistic insight for the whole data set. Thus, using the data from Chen's two studies, they analysed a data set of the flux for 256 compounds. Five of Chen's original data were omitted due to ambiguities in their structures, and the thirteen compounds common to both studies were only included once. Cronin et al. calculated 43 descriptors for each member of the data set including the octanol–water partition coefficient (as log P if available, c log P otherwise, which may have the potential to introduce variance in the study as calculations and predictions of log P often differ—see Chap. 9), topological indices and various measures of hydrogen bonding. Stepwise regression and the removal of outliers considering their residuals produced the following relationship between flux and significant descriptors:

$$\begin{gathered}\log J = -0.561\ \text{HA} - 0.671\ \text{HD} - 0.801\,{}^6\chi - 0.383 \\ [n = 242 \quad r = 0.900 \quad s = 0.464 \quad F = 338]\end{gathered} \tag{5.4}$$

where HA and HD are, respectively, the number of hydrogen bond acceptor and donor groups present on a penetrant, and ${}^6\chi$ is the sixth-order path molecular connectivity.

Thus, the highly significant model describes permeability across the PDMS membrane in terms of hydrogen bonding and, to a lesser extent, molecular topology. The flux is inversely related to the simple count of hydrogen bonding groups available on a molecule, and the topological expression, $^6\chi$, is based on a count of the number of paths of six atoms, irrespective of the presence of heteroatoms and therefore described molecular volume, or molecular bulk. It is, in Eq. 5.4, associated with a decrease in flux as $^6\chi$ increases. Cronin et al. commented that the significance of such a specific descriptor may encode more subtle information on the relative importance of six-membered rings compared to, for example, five-membered rings and their comparative significance in influencing permeation across the PDMS membrane—in a general mechanistic sense, larger or bulkier molecules are less likely to pass across the membrane. In comparing Cronin's model with those developed by Chen, it is clear that Chen's are statistically more significant, which may be due to their analysis of subsets rather than the complete data set. Nevertheless, the models from all three studies do find commonality in that Chen's use of parameters describing molecular charge was rationalised as describing hydrogen bonding, a phenomenon of high significance in Cronin's model. They also found molar solubility in isopropyl alcohol to be significant, and which Cronin also suggested could be related to hydrogen bonding. Cronin also compared their model to the Potts and Guy (1992) algorithm for human skin permeability, highlighting the differences in both models. Nevertheless, solvent selection, particularly after the mechanistic work of Hadgraft, highlighted above, may play a role in producing very different models, as does the comparative simplicity of the PDMS membrane compared to the multilayered and significantly more complex human skin. However, one issue to additionally consider is the limited number of descriptors employed in early QSAR-type studies of human skin, such as Potts and Guy (1992) and Flynn (1990) where permeability was quantified in terms of a small range of descriptors whose significance was determined by reference to experimental studies; the analysis of PDMS might therefore reflect the methodology of analysing a wider range of descriptors; this might also be considered in the significance of $^6\chi$ in Cronin's model, as topological parameters were not calculated by Chen. While this might also speak to the ease with which such parameters can be calculated, particularly by non-experts, it does suggest a limited value in making such comparisons particularly when later QSAR studies of human skin examine a wider range of parameters (e.g. Patel et al. 2002). Further, the composite and possibly covariate nature of parameters such as log P may also lend itself to a more empirical and less mechanistic approach to algorithm development. Thus, studies such as those by Chen et al. (1993, 1996) and Cronin et al. (1998) suggest that more complex methods may be required to discern specific mechanistic information and that the dual purpose of such models—predictive ability and the provision of mechanistic insight—might not always be a relevant outcome for all analyses.

A novel approach was taken to address this issue by applying artificial neural networks (ANNs) (Agatonovic-Kustrin et al. 2001). They used the data originally published by Chen et al. (1993, 1996) and modified by Cronin et al. (1998) for their analysis. They optimised and analysed their neural network model, which was based on a wide range of descriptors similar in type and range to those examined by Cronin et al. They generated a 12-parameter nonlinear QSAR model, based on descriptors that characterise dielectric energy, –OH and $–NH_2–$ groups present on a molecule, the count of ring structures present in a molecule, the lowest unoccupied molecular orbital, EL affinity, molecular weight, total energy, dipole and descriptors of connectivity and molecular bulk. The model they developed indicated that intermolecular interactions (dipole interaction, electron affinity), hydrogen bonding ability (the presence of amino and hydroxyl group) and molecular shape and size (topological shape indices, molecular connectivity indices, ring count) were important for drug penetration through PDMS membranes. log *P* was not found to be a significant descriptor in their analysis, which they suggested was due to the inability of this parameter to account for intramolecular interactions, including intramolecular hydrogen bonding.

As with Cronin's study, Agatonovic-Kustrin et al. found that topological indices were significant. They commented that their inclusion was significant as they could be calculated for any structure, real or hypothetical, and their inclusion was significant for drug discovery and new drug development. Their model included as significant descriptors topological shape indices of the first order (κ^1) and connectivity indices of the first and second order (χ^1 and χ^2, respectively) which allowed specific quantification of molecular shape and bulk properties, describing similarity or dissimilarity of molecules based on the comparative values of the significant topological indices for molecules being compared. Topological shape indices encoded information on structural features such as size, shape, branching pattern, cyclicity and symmetry of molecular graphs. κ values are derived from fragments of one-bond, two-bond and three-bond fragments, with each count being made relative to fragment counts in reference structures. The first-order shape index, κ^1, encodes molecular cycles, with κ^2 and κ^3 encoding linearity and branching, respectively. Thus, the model proposed by Agatonovic-Kustrin et al. shows that an increase in κ^1 decreased membrane permeation due to an increase in molecular size and lipid solubility. χ values indicate the extent of branching present in a molecule, which is the sum of the carbon atoms in a molecule linked to neighbouring carbons atoms, forming the χ index from which specific information on the number of bond fragments can be determined. Such values can be used to quantify aspects of a molecular structure; χ^0, or zero-order connectivity indices, provides information on the number of atoms in a molecule; χ^1, or the first-order connectivity index, encodes the properties of single bonds, being a weighted count of bonds and is related to the types and position of branching in the molecule; and χ^2, the second-order connectivity indices, is derived from fragments of two bond lengths, providing information about types and positioning of branching, indicating structural flexibility. Thus, Agatonovic-Kustrin et al. found that an increase in branching, based on the significance of the χ^1 and χ^2 descriptors in their model, suggested an increase in surface area and molecular volume, resulting in an increased solubility and reduced partition coefficient. Their analysis suggested that the increase in the χ^1 and χ^2 descriptors was consistent with a decrease in membrane penetration and that the χ^1 and χ^2 descriptors were covariant to an extent, although sufficiently different to each encode different, specific characteristics of the penetrating molecules; for example, χ^2 can differentiate between structural isomers, whereas χ^1 values for isomers are identical. Lower values of χ^1 and χ^2 are associated with comparatively more elongated molecules or those with only a single branching atom. They commented that an increase in molecular topology, characterised by the significance of the κ^1, χ^1 and χ^2 descriptors, and an increase in ring count and molecular mass result in a decrease in flux across the PDMS membrane. Thus, mechanistically, a more bulky molecule is less likely to pass through the membrane. Overall, however, the most significant term in their 12-descriptor nonlinear QSAR was dielectric energy—essentially, the change in charge rearrangement of a molecule, which accompanies the change in hydrogen bonding strength. The model proposed by Agatonovic-Kustrin et al. suggested that an increase in dielectric energy is associated with an increase in membrane permeation.

Thus, Agatonovic-Kustrin et al. proposed a highly significant ($r^2 > 0.91$; $RMS_{train} = 0.36$; $RMS_{test} = 0.59$) complex 12-descriptor model which describes the permeation across a PDMS membrane in terms of a wide range of physicochemical descriptors which broadly sit with the model proposed by Cronin et al. (1998). Agatonovic-Kustrin et al. suggest that the specificity and statistical significance of their model can remove the need to conduct laboratory experiments as permeability was not based on experimentally derived parameters.

Geinoz et al. (2002) explored a similar theme with a substantially smaller data set. They characterised the permeability of a model data set across a PDMS membrane for 16 model compounds, and in their analysis, they adjusted for ionisation:

$$f_{\text{ui}} = \frac{1}{(1 + 10^{g})} \tag{5.5}$$

where

f_{ui} is the unionised fraction of the chemical;

g is the relationship between pH and pK; therefore, $g = (\text{pH} - \text{pK}_a)$ for acids and $(\text{pK}_a - \text{pH})$ for bases.

Geinoz et al. developed the following model:

$$\begin{gathered} \log k_p = 0.56 \log P - 0.0108 \sum \text{MHBP}_{\text{do}} - 1.16 \\ [n = 16 \quad r^2 = 0.77 \quad q^2 = 0.61 \quad s = 0.35 \quad F = 21] \end{gathered} \tag{5.6}$$

Thus, their model was very similar to that produced by Cronin et al. (1998) as it related hydrogen bonding (as $\Sigma\text{MHBP}_{\text{do}}$) to permeability. Geinoz et al. did not calculate or model topological descriptors, and while Cronin found such parameters described permeability, Geinoz's model instead saw log P included as a significant descriptor. They compared their model to human skin and commented that it correlated reasonably well ($r^2 = 0.90$) but tended to over-predict. They thus concluded that silicone membranes could provide a useful trend-predictive model for skin penetration.

Ma et al. (2006) developed a QSPR for a PDMS membrane using the heuristic method of mathematical optimisation. Using the Chen/Cronin data sets, they calculated descriptors for each molecule using Comprehensive Descriptors for Structural and Statistical Analysis (CODESSA) software. The heuristic method was used to select descriptors and to develop their linear QSAR. A highly significant (r^2 = 0.844; RMSE = 0.438) 4-descriptor model was proposed, where the significant terms were the count of hydrogen bond acceptor sites on a molecule, the gravitation index, H-donors charged surface area and the weighted positive-charged partial surface area. This study is similar in many respects to those described above (Chen et al. 1993, 1996; Cronin et al. 1998; Agatonovic-Kustrin et al. 2001) in that it described permeability across a PDMS membrane in terms of similar molecular features which appear to relate to broader molecular phenomena, such as hydrogen bonding. In most of these studies, similar data sets are used which produce different outputs depending on the method of analysis used. The specific detail of each model, and the specific descriptors returned as significant in each study, perhaps reflects the difficulty of modelling such experimental data in such specific ways and suggests the need to present the output from such models in a simplified, consistent manner as it is otherwise difficult to ascertain the significance of such specific molecular analysis in the required mechanistic context of bulk partition and permeation into and across a membrane.

Several other studies have focused on developing quantitative expressions of permeability of penetrants into and across PDMS, or related, membranes. Wasdo et al. (2008) modelled flux across silicone membranes from aqueous solutions, fitting their data to the Roberts–Sloan or modified Kasting–Smith–Cooper models for a series of prodrugs, suggesting that the Roberts–Sloan model gave a better fit to that database, as well as to data sets relating maximum flux from water across mouse and human skin. Kang et al. (2007) also used PDMS membranes to consider a formulation-based model for assessing the enhancement effects of a range of terpenes. New membrane types are also being reported, with the aim to produce a hybrid lipophilic—hydrophilic membrane that is more representative of the heterogeneity of mammaliam skin, and artificial membranes are finding application in high-throughput models for skin permeability (i.e. Ottaviani et al. 2006, 2007).

Several studies are working towards building relationships between human skin permeability and permeability across skin from other relevant mammals, as well as PDMS and related membranes (Wasdo et al. 2009; Sugibayashi et al. 2010).

Nevertheless, there is an obvious paucity of QSAR analyses of PDMS permeability, particularly compared to similar studies for human skin. Despite clear reasons for using PDMS experimentally, as highlighted by the work of Hadgraft and others (described above) with a number of viable models of human skin permeability, and in the context of regulatory approval for new pharmaceutical formulations, it is clear that the interest in, and application of, QSPRs for PDMS membranes is of limited value. This is highlighted somewhat by Moss et al. (2011) who produced a series of machine learning models for permeability across a number of membranes, including PDMS. Their study, which is described in detail in Chap. 7, highlighted the issues associated with quality of input data, demonstrating that model quality was significantly influenced by the availability and quality of data. In doing so, they showed poor relationships between permeability models for mammalian skin permeability and artificial membranes, including the PDMS membrane. Nevertheless, the potential benefits in developing a model of permeability for a PDMS membrane is enormous, including optimisation of permeant selection and design in topical and transdermal drug delivery, which could potentially offer a significant reduction in the number of animals used currently in such studies.

References

Agatonovic-Kustrin S, Beresford R, Pauzi A, Yusof M (2001) ANN modelling of the penetration across a polydimethylsiloxane membrane from theoretically derived molecular descriptors. J Pharm Biomed Anal 26:241–254

Ahmed M, Hadgraft J, Kellaway IW (1983) Phenothiazine transport across liquid–lipid, phospholipid and soft polymer membranes. 2. Int J Pharm 13:227–237

Caussin J, Gooris GS, Janssens M, Bouwstra JA (2008) Lipid organization in human and porcine stratum corneum differs widely, while lipid mixtures with porcine ceramides model human stratum corneum lipid organization very closely. Biochim Biophys Acta 6:1472–1482

Chen Y, Vayumhausuwan P, Matheson LE (1996) Prediction of flux through polydimethylsiloxane membranes using atomic charge calculations: application to an extended data set. Int J Pharm 137:149–158

Chen Y, Yang WL, Matheson LE (1993) Prediction of flux through polydimethylsiloxane membranes using atomic charge calculations. Int J Pharm 94:81–88

Cornwell PA, Barry BW (1995) Effects of penetration enhancer treatment on the statistical distribution of human skin permeabilities. Int J Pharm 117:101–112

Cronin MTD, Dearden JC, Gupta R, Moss GP (1998) An investigation of the mechanism of flux across polydimethylsiloxane membranes by the use of quantitative structure-permeability relationships. J Pharm Pharmacol 50:143–152

Dias M, Hadgraft J, Lane ME (2007) Influence of membrane-solvent-solute interactions on solute permeation in model systems. Int J Pharm 336:108–114

Fasano WJ, Manning LA, Green JW (2002) Rapid integrity assessment of rat and human epidermal membranes for in vitro dermal regulatory testing: correlation of electrical resistance with tritiated water permeability. Toxicol In Vitro 16:731–740

Feldstein MM, Raigorodskii IM, Iordanskii AL, Hadgraft J (1998) Modelling of percutaneous drug transport in vitro using skin-imitating Carbosil membrane. J Cont Rel 52:25–40

Flynn GL (1990) Physicochemical determinants of skin absorption. In: Gerrity TR, Henry CJ (eds) Principles of route-to-route extrapolation for risk assessment. Elsevier, New York, pp 93–127

Frum Y, Eccleston GM, Median VM (2007) Evidence that drug flux across synthetic membranes is described by normally distributed permeability coefficients. Eur J Pharm Biopharm 67:434–439

Geinoz S, Rey S, Boss G, Bunge AL, Guy RH, Carrupt PA, Reist M, Testa B (2002) Quantitative structure-permeation relationships for solute transport across silicone membranes. Pharm Res 19:1622–1629

Gullick DR, Pugh WJ, Ingram MJ, Cox PA, Moss GP (2010) Formulation and characterization of a captopril ethyl ester drug-in-adhesive-type patch for percutaneous absorption. Drug Dev Ind Pharm 36:926–932

Kang L, Yap CW, Lim PFC, Chen YZ, Ho PC, Chan YW, Wong GP, Chan SY (2007) Formulation development of transdermal dosage forms: quantitative structure-activity relationship model for predicting activities of terpenes that enhance drug penetration through human skin. J Cont Rel 120:211–219

Kasting GB, Francis WR, Filloon TG, Meredith MP (1992) Improving the sensitivity of in vitro skin penetration studies. AAPS Annual Meeting, San Antonio, TX

Khan GM, Frum Y, Sarheed O, Eccleston GM, Median VM (2005) Assessment of drug permeability distributions in two different model skins. Int J Pharm 303:81–87

Klang V, Schwarz JC, Lenobel B, Nadj M, Aubock J, Wolzt M, Valenta C (2012) In vitro vs. in vivo tape stripping: validation of the porcine ear model and penetration assessment of novel sucrose stearate emulsions. Eur J Pharm Biopharm 3:604–614

Kuchler S, Struver K, Friess W (2013) Reconstructed skin models as emerging tools for drug absorption studies. Exp Op Drug Met Tox 9:1255–1263

Kumprakob U, Kawakami J, Adachi I (2005) Permeation enhancement of ketoprofen using a supersaturated system with antinucleant polymers. Bio Pharm Bull 28:1684–1688

Ley EE, Bunge AL (2007) Chemical transport in silicone rubber membranes from pure powders and saturated aqueous solutions. J Memb Sci 292:35–44

Liu P, Nightingale J, Kurihara-Bergstrom T (1991) Variation of in vitro skin permeation data for ionic compounds. Pharm Res 8:S137

Ma WP, Luan F, Zhao CY, Zhang XY (2006) QSAR prediction of the penetration of drugs across a polydimethylsiloxane membrane. QSAR Comb Sci 25:895–904

McAuley WJ, Mader KT, Tetteh J, Lane ME, Hadgraft J (2009) Simultaneous monitoring of drug and solvent diffusion across a model membrane using ATR_FTIR spectroscopy. Eur J Pharm Sci 38:378–383

McAuley WJ, Olivera G, Mohammed D, Beezer AE, Hadgraft J, Lane ME (2010) Thermodynamic considerations of solvent/enhancer uptake into a model membrane. Int J Pharm 396:134–139

Moss GP, Gullick DR, Cox PA, Alexander C, Ingram MJ, Smart JD, Pugh WJ (2006) Design, synthesis and characterisation of captopril prodrugs for enhanced percutaneous absorption. J Pharm Pharmacol 58:167–177

Moss GP, Sun Y, Wilkinson SC, Davey N, Adams R, Martin GP, Prapopoulou M, Brown MB (2011) The application and limitations of mathematical modelling in the prediction of permeability across mammalian skin and polydimethylsiloxane membranes. J Pharm Pharmacol 63:1411–1427

Netzlaff F, Lehr CM, Wertz PW, Schaefer UF (2005) The human epidermis models EpiSkin, SkinEthic and EpiDerm: an evaluation of morphology and their suitability for testing phototoxicity, irritancy, corrosivity, and substance transport. Eur J Pharm Biopharm 2:167–178

Olivera G, Beezer AE, Hadgraft J, Lane ME (2010) Alcohol enhanced permeation in model membranes. Part 1. Thermodynamic and kinetic analyses of membrane permeation. Int J Pharm 393:61–67

Ottaviani G, Martel S, Carrupt PA (2007) In silico and in vitro filters for the fast estimation of skin permeation and distribution of new chemical entities. J Med Chem 50:742–748

Ottaviani G, Martel S, Carrupt PA (2006) Parallel artificial membrane permeability assay: a new membrane for the fast prediction of passive human skin permeability. J Med Chem 49:3948–3954

Patel H, ten Berge W, Cronin MTD (2002) Quantitative structure-activity relationships (QSARs) for prediction of skin permeation of exogenous chemicals. Chemosphere 48:603–613

Pellett MA, Waktinson AC, Hadgraft J, Brain KR (1994) Diffusion of deuterated ethanol in synthetic membranes using ATR-FTIR spectroscopy. Eur J Pharm Sci 2:155

Petitot F, Frelon S, Moreels AM, Claraz M, Delissen O, Tourlonias E, Dhieux B, Maubert C, Paquet F (2007) Incorporation and distribution of uranium in rats after a contamination on intact or wounded skin. Health Phys 5:464–474

Potts RO, Guy RH (1992) Predicting skin permeability. Pharm Res 9:663–669

Roper CS, Simpson AG, Madden S, Cameron BD (2000) Tritiated water permeability coefficient assessment and rejection criteria for barrier function of human skin. In: Brain KR, Walters KA (eds) Perspectives in percutaneous penetration, vol 7A. STS Publishing, Cardiff, UK

Schafer-Korting M, Bock U, Diembeck W, Dusing HJ, Gamer A, Haltner-Ukomadu E, Hoffman C, Kaca M, Kamp H, Kersen S, Kietzmann M, Korting HC, Krachter HU, Lehr CM, Liebsch M, Mehling A, Muller-Goymann C, Netzlaff F, Niedorf F, Rubbelke MH, Schaefer U, Schmidt E, Schreiber S, Spielmann H, Vuia A, Weimer M (2008) The use of reconstructed human epidermis for skin absorption testing: results of the validation study. Alt Lab Anim 2:161–187

Shumilov M, Touitou E, Godin B, Ainbinder D, Yosha I, Tsahor-Ohayon H (2009) Evaluation of a polysiloxane-collagen biphasic membrane: a model for in vitro skin permeation studies. J Drug Del Sci Tech 19:289–294

Sugibayashi K, Todo H, Oshizaka T, Owada Y (2010) Mathematical model to predict skin concentration of drugs: toward utilization of silicone membrane to predict skin concentration of drugs as an animal testing alternative. Pharm Res 27:134–142

Vallet V, Cruz C, Josse D, Bazire A, Lallement G, Boudry I (2007) In vitro percutaneous penetration of organophosphorus compounds using full-thickness and split-thickness pig and human skin. Toxicol In Vitro 6:1182–1190

Van Gele M, Geusens B, Brochez L, Speeckaert R, Lambert J (2011) Three-dimensional skin models as tools for transdermal drug delivery: challenges and limitations. Exp Op Drug Del 8:705–720

Waktinson AC, Hadgraft J, Walters KA, Brain KR (1994) Measurement of diffusional parameters in membranes using ATR-FTIR spectroscopy. Int J Cos Sci 16:199–210

Wasdo S, Juntunen J, Devarajan H, Murray T, Nickels D, Singh S, Shanks T, Ulmer K, Sloan KB (2008) Modelling of flux through silicone membranes from water. Eur J Pharm Sci 34:321–332

Wasdo SC, Juntunen J, Devarajan H, Sloan KB (2009) A correlation of flux through a silicone membrane with flux through hairless mouse skin and human skin in vitro. Int J Pharm 373:62067

Watkinson AC, Brain KR, Walters KA, Grabarz RS, Sharma RK (1998) Is it logical to reject skin samples based on water permeability data? In: Brain KR, Walters KA (eds) Perspectives in percutaneous penetration, vol 6A. STS Publishing, Cardiff, UK, p 62

Wenkers BP, Lippold BC (1999) The statistical distribution of in vivo skin permeability. Pharm Pharmacol Commun 5:425–427

Williams AC, Cornwell PA, Barry BW (1992) On the non-Gaussian distribution of human skin permeabilities. Int J Pharm 86:69–77

Woolfson AD, McCafferty DF, McClelland KH, Boston V (1988) Concentration-response analysis of percutaneous local anaesthetic formulations. Br J Anaesth 61:589–592

Woolfson AD, McCafferty DF, Moss GP (1998) Development and characterisation of a moisture-activated bioadhesive drug delivery system for percutaneous local anaesthesia. Int J Pharm 169:83–94

Woolfson AD, McCafferty DF (1993) Percutaneous local anaesthesia. Ellis Horwood, London

Chapter 6
Other Approaches to Modelling Percutaneous Absorption

The preceding chapters of this book have dealt with the generalised field of models for percutaneous absorption which are, by and large, based either on the use of Flynn's data set (Flynn 1990) or on the variations thereon, using subsets of data sets which reflect specific types of molecules and which are generally analysed by rudimentary statistical approaches—mostly multiple linear regression analysis or similar methods.

While such approaches might present themselves as a large and important body of work, approaching almost a consensus, it clearly does not reflect the breadth of research in this field and the range of other methods which have been applied to this problem domain. Thus, the next three chapters will address various aspects of the field which are not addressed by the general models of skin absorption. Some of this work has been presented in isolated studies, and the reasons why such studies have not been further developed will be addressed later. Examples include the use of methods which have found sporadic use, or which use different endpoints, such as a number of studies by Roberts and colleagues over the last ten years or so which focus not on permeability but transdermal flux, and which are discussed below.

One very good example of the need for different models again begins by reflecting on Flynn's approach. This is based on the rationale that penetrants will most likely be absorbed into the skin from saturated aqueous solutions. This also infers that an infinite dose is applied to the skin. Clearly, whilst representative of a great many exposure or dosing scenarios, there are situations when such models do not apply. This may include, for example, systems where an exposure may occur from non-aqueous or a volatile solvent, or from a sub-saturated (i.e. finite) dose exposure. Models of non-steady-state and finite-dose experiments will be considered in Chap. 8.

Thus, this chapter will aim to consider and, where relevant, collate those models that do not fit the classifications discussed in Chap. 4. Further, recent models with relevance to the cosmetic sciences, such as those proposed by Gregoire et al. (2009), will be considered.

As discussed previously, several published QSAR-type models sit somewhat outside the mainstream. These studies are often characterised by their application of conventional methodology to specific data sets. For example, Le and Lippold (1995) used a data set of four nicotinic acid esters, finding a relationship between

G.P. Moss et al., *Predictive Methods in Percutaneous Absorption*,
DOI 10.1007/978-3-662-47371-9_6

lipophilicity and permeability in a guinea pig model for these four molecularly similar chemicals (see Chap. 4). Diez-Sales et al. (1993), using a rat skin model with and without the dermal layer, assessed the permeability of a series of 4-alkylanilines. Interestingly, they found different trends depending on the tissue used; while correlations were generally bilinear in the absence of the dermal layer, they tended towards a hyperbolic relationship between permeability and significant physicochemical descriptors, particularly lipophilicity. They suggested that the often-observed heterogeneity in the skin should be attributed to the epidermal and dermal layers, rather than being solely attributed to the *stratum corneum*, and that this proposal was common to a range of mammalian species (rat, mouse, human). Nevertheless, their results and findings, particularly in terms of developing quantitative models and relationships between permeability and physicochemical parameters, should be considered in the context of the penetrants examined, and therefore the size and diversity of their data set.

Most studies using variations on the methods of Flynn focus on permeability from saturated aqueous solutions. Dal Pozzo et al. (1991), however, examined the permeability of a series of esters, which were applied to the skin as saturated solutions or pure liquids. They observed a plateau in permeability which was related to lipophilicity but commented that this was due to the effect of water when it was used as the solvent in the donor compartment of the diffusion experiment. Despite the comparatively small data set, this clearly suggests that the solvent choice may limit the applicability of models and any inferences from them.

Donor solubility was also investigated by Bast (1997), who looked at the influence of solubility, and permeant size, on skin absorption in a rabbit model. Bast found that, with the application of exogenous chemicals to the skin in lipophilic vehicles, there was a significant decrease in the permeability coefficient, something which was addressed more qualitatively, and with greater clinical emphasis, by McCafferty and Woolfson (1993) for a single penetrant (amethocaine) and which was used to formulate a clinically viable formulation strategy—the amethocaine phase-change system (McCafferty et al. 2000). This was associated primarily with a permeant lipophilicity, as represented by log P, of 3.0–3.5. Similar findings were found for increases in permeant molecular weight, suggesting a degree of covariance between these physicochemical descriptors. In general, while such studies clearly have a use, it is often confined to certain types of chemicals or homologous series and often have little use outside such a confined molecular space. In many cases, these studies also use different methods, or may be formulation specific, and thus cannot be added to the Flynn (1990) data set to expand its membership. However, even with relatively small data sets, studies can offer a wider context. For example, the study by Morimoto et al. (1992), which is discussed in Chap. 4, employs a comparatively small data set ($n = 16$), but its contents are structurally diverse, and thus, its findings—in particular, its reporting of a biphasic relationship between permeability and physicochemical descriptors—may have a broader context.

Nevertheless, despite the limitations of models based on small data sets, they can also offer insight into other processes. A significant example of this is the study by Borras-Blasco et al. (2004). Building on earlier work (e.g. Borras-Blasco et al. 1997, which proposed empirical relationships between the effect of skin penetration enhancers and the physicochemical properties of penetrants), they used a mathematical approach to estimate the influence of sodium lauryl sulphate (SLS, at concentrations from 0.24 to 5 % w/w) on the permeation of seven model drugs with a wide range of lipophilicities, from −0.95 to 4.21. They initially found that the experimental method employed was important to consider and that it was related to the log P of permeants. Specifically, pretreatment of the skin used in their in vitro experiments did not affect the permeability (measured as k_p, the permeability coefficient) of model drugs where log $P > 3.0$. However, where log $P < 3.0$ increases in permeability were observed which were dependent on the concentration of SLS applied to the skin and the lipophilicity of the compounds tested. Thus, a hyperbolic equation was proposed which related the inverse of the ability of SLS to act as an enhancer (1/ER, where ER is the enhancement ratio for permeability of each model drug under the different experimental conditions used, which was based on the approach proposed by Williams and Barry (1991) where ER was a function of the permeability before and after the application of the penetration enhancer):

$$\frac{1}{\mathrm{ER}} = \frac{P}{18.44 \cdot C - 3.76 + P} \tag{6.1}$$

where

P is the partition coefficient of the permeant between the membrane and the donor vehicle;

C is the concentration (in this case, the solubility) of the permeant in the donor solution.

Validation of this model produced excellent fits between experimental and predicted permeabilities ($r^2 > 0.94$):

$$\frac{1}{\mathrm{ER}_{\mathrm{theoretical}}} = 1.04 \cdot \frac{1}{\mathrm{ER}_{\mathrm{experimental}}} - 0.068 \tag{6.2}$$

They also applied their approach to previously published data (by Diez-Sales et al. 1996) and found a significant fit to a linear model, which was similar to Eq. 6.2, for the skin penetration enhancer Azone®, which is known to enhance permeation based on the lipophilicity of the permeant:

$$\frac{1}{\mathrm{ER}_{\mathrm{theoretical}}} = 0.86 \cdot \frac{1}{\mathrm{ER}_{\mathrm{experimental}}} - 0.08 \quad [r^2 > 0.77] \tag{6.3}$$

where 1/ER was, for this data set, found to be:

$$\frac{1}{\mathrm{ER}} = \frac{P}{2.83 \cdot C - 4.37 + P} \tag{6.4}$$

They commented that, despite significant chemical differences between SLS and Azone®, particularly in their hydrophilic domains which might imply different interactions with the *stratum corneum*, in a qualitative sense—and in the context of the nature and size of the data set used for this study—their effects on skin permeability were very similar and could be predicted very well by the models proposed (Eqs. 6.1–6.4).

In such a context of model range and/or limitation, particularly in the context of vehicles more complex than those normally associated with Flynn-based permeability models, Gregoire et al. (2009) addressed a significant issue in the development of predictive algorithms of skin permeation—the lack of applicability to a wider range of vehicles. They developed a predictive model which estimated the cumulative mass of a chemical absorbed into and across the skin from topical formulations (i.e. cosmetic or dermatological preparations). In doing so, they assumed that a steady state was achieved despite the application of a finite dose, that vehicle effects were small relative to the precision (or otherwise) of the prediction and that each formulation could be treated as an oil-in-water emulsion in which only the aqueous fraction of the chemical was available for permeation into the *stratum corneum*. In analysing a data set of 101 ex vivo human skin experiments for 36 chemicals they found that, in most cases, the difference between experimental and predicted permeability was less than fivefold and that the model was able to accurately estimate permeation for two chemicals not in the data set. Nevertheless, their model highlights the complex issues associated with predicting the permeability of exogenous chemicals from a range of formulations and, in doing so, addresses the limitations of current models—which focus mostly on saturated aqueous solutions and highlights the challenges ahead in this field. This is an issue that has, more broadly, been discussed by others (Selzer et al. 2013) and which will be examined in more detail in subsequent chapters.

Another theme touched on by Flynn (1990) was the nature of models which were not "global" in the sense that they were represented by a single algorithm. An elegant example of how this approach has been taken forward is Mitragotri's (2003) discussion, in the context of a porous pathway approach, of multiple permeation pathways based on permeant physicochemical properties. This approach attempts to discuss the permeation of hydrophilic molecules as, in general, permeation of hydrophobic molecules is reasonably well described by lipid-based models. Several models have described this approach as a "porous pathway" model which properly accounts for the permeation of hydrophilic permeants. For example, Peck et al. (1994) introduced the concept of "hindered diffusion" of polar molecules through the skin by examining a small group of model, hydrophilic, compounds (urea, mannitol, sucrose and raffinose) and describing their permeation. This approach has been developed in other studies (e.g. Hatanaka et al. 1990; Kim et al. 1992;

Morimoto et al. 1992; Lai and Roberts 1998, 1999). Thus, the general expression for permeability via the porous pathway is given by:

$$k_p = \frac{\varepsilon D_p^{\text{pore}}}{\tau \Delta x} \tag{6.5}$$

where

ε, τ and Δx are the porosity, tortuosity and thickness of the membrane, respectively;

D_p^{pore} is the diffusion coefficient of the permeant in the liquid-filled pores of the membrane.

The "hindered diffusion" model considers that D_p^{pore} is a function both of the membrane and the permeant, being dependent upon pore size (radii) and the diffusion coefficient of the permeant at infinite dilution. Despite being able to clearly characterise porosity and its influence on the permeation of highly hydrophilic permeants, such models have found limited application due mainly to the lack of defined links between pore radii and aspects of skin morphology, such as pore density.

Thus, Mitragotri (2003) approached this problem by examining solute permeation through four possible routes in the *stratum corneum*: free volume diffusion through lipid bilayers (using scaled particle theory), lateral diffusion along lipid bilayers (determined from literature data), diffusion through pores (from the "hindered transport" theory) and diffusion through shunts (via the application of a simple diffusion model). Mitragotri's analysis resulted in a series of models which described each pathway. Solute permeability across the *stratum corneum* for hydrophobic solutes was described by the expression:

$$K_p^{fv}(r, K_{o/w})(\text{cm/s}) = 5.6 \times 10^{-6} \cdot K_{o/w}^{0.7} \exp(-0.46\, r^2) \tag{6.6}$$

where r is the radius (units Å), which can be calculated as described by van der Bondi (1964) or approximated from the molecular weight (MW) of the penetrant, based on the relationship $4/3\pi r^2 = 0.9087\,\text{MW}$ (Mitragotri et al. 1999).

They also proposed a method to estimate the lateral diffusion of lipids, proposing that the diffusion of large solutes that are incorporated into the bilayer is related to the lateral diffusion coefficients of lipid molecules:

$$K_P^{\text{lateral}} = \frac{K_{o/w}^{0.7}}{3.6} \cdot D_b^{\text{lipid}} \tag{6.7}$$

Equation 6.7 can, in the context of several assumptions (i.e. consideration of $r = 4.3$ Å, where $D_b \approx D_0 \exp(-Ar^2)$ which is comparable to lateral lipid diffusion in other systems), be rewritten as:

$$K_p^{\text{lateral}}(\text{cm/s}) = K_p^{\text{lateral}}\left(K_{o/w}\right) = 8 \times 10^{-10} K_{o/w}^{0.7} \tag{6.8}$$

Mitragotri highlighted that lateral lipid diffusion is not dependent on permeant size and that the transition from free volume diffusion to lateral diffusion occurs at a radius of approximately 4.3 Å, which corresponds to a value of $r^2 = 18.5$, or an approximate molecule weight of 380 Da.

Transport through pores was assumed to play a major role in the permeation of hydrophilic permeants. Based on the assumption that polar or aqueous pathways—often considered as "pore" pathways—exist and will favour the permeation of hydrophilic molecules (Cornwell and Barry 1993; Edwards and Langer 1994; Menon and Elias 1997), Mitragotri placed such a pathway in the context of lipid bilayer imperfections which may be observed as grain boundaries, lattice vacancies, defects in lattice structures or any combination of such features, and which may provide a "polar" pathway for the permeation of hydrophilic molecules. Equation 6.5 shows the general expression for permeability of a solute through a porous membrane, from which the hindrance factor may be determined, following the method of Deen (1987):

$$\begin{array}{l} H(\lambda) = (1-\lambda)^2 \cdot (1 - 2.104\lambda + 2.09\lambda^3 - 0.95\lambda^5 \\ \text{[for low molecular weight solutes, where } \lambda < 0.4] \end{array} \tag{6.9}$$

where λ is the ratio of the hydrodynamic radius of the permeant and the effective pore radius of the membrane.

And

$$H(\lambda) = \frac{6\pi}{\frac{9}{4}\pi^2\sqrt{2(1-\lambda)^{-5/2}\left[1+\sum_{n=1}^{2} a_n(1-\lambda)^n\right] + \sum_{n=0}^{4}(a_{n+3})\lambda^n}} \tag{6.10}$$

$$\text{[for low molecular weight solutes, where } \lambda < 0.4]$$

where $a_1 = -1.217$, $a_2 = 1.534$, $a_3 = -22.51$, $a_4 = -5.612$, $a_5 = -0.3363$, $a_6 = -1.216$ and $a_7 = 1.647$

Thus, to estimate permeability via this pathway ε, τ, and Δx, the porosity, tortuosity and thickness of the membrane, respectively, need to be known. Finally, the last part of the four-compartment model, which considers transport through skin appendages, may ultimately be represented by the expression:

$$K_p^{\text{shunt}}(\text{cm/s}) = 2 \times 10^{-9} \tag{6.11}$$

Mitragotri commented that this route is only of significance for the permeation of large (MW > 100,000 Da) hydrophilic molecules.

Thus, this specific and comprehensive model compares well with experimental results in the data sets used in this study. In addition, different permeants have different weightings for each pathway based on their physicochemical properties. Using the Johnson-modified Flynn data set (Flynn 1990; Johnson et al. 1995; $n = 83$), they found excellent correlations between measured and estimated permeability, with a mean error of approximately 6 %. They found that the contribution of free volume diffusion decreases exponentially and is dominant for small permeants (less than 4 Å), which is related to the radii of the pores and the solutes. Skin permeability to hydrophobic solutes exhibits significant size selectivity, whereas the contribution of lateral lipid diffusion was considered to be significant for larger solutes but to not be characterised by size dependence. The transport of hydrophilic drugs is hypothesised to occur through pores in the *stratum corneum* lipid bilayers, which may be the result of structural imperfections within the bilayer. Transport through such "pores" is characterised by porosity, tortuosity and pore size distribution. Pore size and porosity are characteristics associated entirely with the skin morphology, whereas tortuosity depends on the *stratum corneum* structure as well as the solute size.

Different solutes were shown to differ in the relative contributions each pathway makes to their overall permeability, and the contribution of each pathway to skin permeability is a function of size and lipophilicity. Thus, permeation of small, hydrophobic solutes is mostly via free volume diffusion. As solute size increases, the free volume pathway diminishes to insignificance and permeability is defined predominately by lateral lipid diffusion. For highly hydrophilic solutes where, for example, $K_{o/w} < 0.01$, skin permeability is a function of pore and shunt permeability. Finally, for permeants of moderate hydrophilicity ($K_{o/w} \sim 0.01–1$), permeability is related mostly to free volume diffusion through lipid bilayers.

For further details of this excellent study, the reader is referred to Mitragotri's excellent paper (Mitragotri 2003).

One of the most original, and important, models that sit outside the context of Potts- and Guy-type algorithms based on the permeability coefficient was reported by Magnusson et al. (2004). They commented that the delivery rate at which the solute is absorbed into and across the skin is highly significant in terms of systemic and local therapeutic or toxicological endpoints. More so than the permeability coefficient (k_p) as, in practice, the maximum flux (J_{max}, with units of amount/time/surface area), usually at steady-state, is of most interest in determining the maximum absorption.

Very few studies have therefore estimated skin permeability using flux. For example, Higuchi and Davis (1970) described a simple modelling approach that allowed rational way to predict the degree of lipophilicity which would result in maximal permeation. Kasting et al. (1987) found a relationship between flux (as $\log J_{max}$) and both solubility in octanol and molecular volume for 35 chemicals administered to the human skin from saturated propylene glycol solutions. Roberts and Sloan (2000) also predicted the flux of a series of prodrugs ($n = 41$) of 5-fluorouracil, theophylline and 6-mercaptopurine using models with a number of approaches which considered separate paths for lipid and aqueous permeation in parallel and for both pathways in series. They found that flux was related to descriptors of lipophilicity and molecular weight. Excellent ($r^2 > 0.9$ for all models derived) correlations between predicted and measured permeabilities were found, which compared at least as well to a modified version of the Potts and Guy equation. Their solvatochromic series/parallel model provided the best fit and suggested that it provided further support against theories of a high-capacity aqueous-only pathway across the skin, as well as providing insight into how drugs should be modified to maximise permeation. They were also able to differentiate their models based on penetrant lipophilicity, with a lipid-aqueous in-series pathway model best describing permeability for penetrants where log P was less than 0.8, and a lipid-only path describing permeation for penetrants where log P was greater than 1.0.

It is interesting to note that, of three models which used flux and not permeability to model permeation (Kasting et al. 1987; Cronin et al. 1998; Roberts and Sloan 2000), they all used non-aqueous solvents and, while offering significant findings in terms of solubility effects and their influence on the permeability process, they do limit extrapolation of their findings to other, more widely examined systems.

Nevertheless, the vast majority of studies which model mathematically the process of skin permeability, and which have been described in the previous chapters, do so from aqueous solutions and record their output as the permeability coefficient, k_p (cm/s or ch/h) which is essentially a concentration-corrected adjustment of the flux. The flux of any solute at a given concentration may be defined as the product of maximum steady-state flux and the fractional solubility of the potential penetrant in that formulation. Thus, if the maximal flux is known for a particular solute, its flux from any vehicle can be estimated using its fractional solubility in the vehicle once potential changes in the skin barrier function are considered (Roberts et al. 2002).

Thus, Magnusson et al. (2004) collated the available literature for human skin permeation and aimed to develop a global model which defined the relationship between flux (as J_{max}) and the physicochemical properties of the solutes contained in their data set. In an extensive experimental design, they developed a series of models based on a range of conditions: for full- and split-thickness skin, ionised solutes, pure solutes and maximum fluxes from propylene glycol (the last two of which were used for validation only as they may affect skin condition). Stepwise regression indicated that, for their training set, molecular weight was the main predictor of log J_{max}:

$$\begin{gathered}\log J_{\text{max}} = -3.90 - 0.0190\ \text{MW} \\ [n = 87\ r^2 = 0.847\ p < 0.001]\end{gathered} \tag{6.12}$$

Experimental temperature dependence (as MW/T) did not substantially improve the model ($r^2 = 0.850$) which the authors suggested was obscured by experimental variance due to the multiple sources of their data set. Inclusion of log S_{oc}, after Kasting et al. (1987) improved the model slightly (to an r^2 of 0.856) suggesting that Kasting's "free volume" model for diffusion of solutes within *stratum corneum* lipids, is one contributor for a dependency of flux on molecular size. Addition of further descriptors other than molecular weight (melting point, increased r^2 0.879; melting point and hydrogen bond acceptor ability, increased r^2 to 0.917) to the model resulted in marginal increases in model quality, the latter of which is significant. Analysis of their full, collated data set resulted in the following algorithm:

$$\begin{aligned} \log J_{\max} &= -4.52 - 0.0141\,\mathrm{MW} \\ [n &= 278\ r^2 = 0.688\ p < 0.001] \end{aligned} \tag{6.13}$$

As with Eq. 6.1, increases in model quality were observed when other descriptors (melting point and hydrogen bond acceptor groups) were added to the analysis. The authors suggested that molecular weight can be used to give an initial estimate for $J_{\max}$ for any solute in a saturated aqueous solution or as a pure solute. Departures from this model may be due to the effects certain penetrants can exert on skin permeability and that such effects may be modelled, and therefore used to correct the main model, by consideration of enhancer–solvent property relationships. The authors also comment, from Singh and Roberts (1996) for example, that molecular weight is the only significant determinant of blood clearance. Therefore, application of their model to an in vivo situation where the dermal capillary bed lies just below the dermo-epidermal junction indicates that consideration of dermal resistance was unnecessary to model in vivo predictions, suggesting that molecular weight is the key determinant to systemic uptake irrespective of whether the rate-limiting step for skin absorption is partition into and across the *stratum corneum* or removal of the penetrant from local tissue via the dermal vasculature. Thus, they concluded that molecular weight is the main predictor for flux across ex vivo human skin and that predictions could be marginally improved by the inclusion of experimental temperature (as MW/T), log S_{oc}, the count of hydrogen bond acceptors on a potential penetrant and melting point. Their model also predicted well permeation through their other data subsets (for full- and split-thickness skin and for pure solutes, ionised drugs and for flux from saturated propylene glycol solutions).

This work has been expanded upon by Zhang et al. (2009), who investigated the mechanistic dependence of maximum flux on other solute physicochemical parameters. In doing so, they emphasised the significance of flux which, for a given penetrant, is thermodynamically invariant in describing the penetration process, whereas the permeability coefficient is dependent on the formulation applied. Using a data set of ten phenols with similar molecular weights and hydrogen bonding properties but differing lipophilicities, they measured maximum flux through human epidermal tissue. They reported a bilinear, or Gaussian, relationship between flux and lipophilicity (as log P) with its maximum between log P values of 2.7–3.1. Lag times and diffusivities were predominately independent of lipophilicity. The trends observed in *stratum corneum* fluxes with changing lipophilicities were attributed by the authors to variations in *stratum corneum* solubility rather than from diffusional or partitioning barrier effects at the interface of the *stratum corneum* and the viable epidermis. Thus, the solute solubility in the *stratum corneum*, S_{SC}, rather than diffusional resistance in deeper skin tissues due to aqueous boundary layers (or an inability for a solute to partition into a receptor fluid in an ex vivo study), is responsible for the parabolic–Gaussian behaviour observed for their data set. The observed trend indicates a decrease in flux as lipophilicity increases (above a log *P* of approximately 3) for solutes of a similar molecular size. This is attributed to partition rate-limited permeation for less water-soluble solutes. Thus, they define flux as being dependent on partitioning, which is related to lipophilicity, and diffusivity, which is related to solute size and hydrogen bonding, an observation which is consistent with their experimental findings.

Zhang and colleagues subsequently explored their findings in more detail (Zhang et al. 2011). They contextualised their study with the principle that the maximum skin flux of solutes is unaffected by its vehicle *unless* the vehicle exerts an effect on the nature of the skin barrier. They therefore examined how the use of cosolvent systems commonly attributed to being enhancers of skin penetration influence the maximum fluxes of their model penetrants. As in their previous study (Zhang et al. 2009), they used as a data set ten phenolic compounds of similar molecular weight and hydrogen bonding properties but different lipophilicities. The same data set was employed in their second study, but a range of solvent systems were used (60 % propylene glycol/40 % water; 40 % propylene glycol/60 % water; 100 % water). They found that maximum flux and solubility within the *stratum corneum* increased as the amount of propylene glycol in the solvent system was increased, but that diffusivity was independent of the solvent composition; thus, the increase in flux was attributed to *stratum corneum* solubility, which is vehicle dependent. Further, the solubility in the *stratum corneum* depended on the ability of different formulations to penetrate to different extents into the *stratum corneum* and the amount of each compound dissolved in a particular solvent system. Further detailed mechanistic insight was provided by infrared spectroscopy and multiphoton microscopy studies, which indicated that, for the model penetrant β-naphthol, increased uptake into the *stratum corneum* was due to an increased solubility of the penetrant in the intercellular lipids of the *stratum corneum*; thus, the use of propylene glycol was able to increase flux into and across the skin for similar-size molecules. A similar diffusivity was found for all compounds and was independent of the penetrant size or the nature of the vehicle used. As in their earlier study, they again concluded that the maximum flux was found for chemicals with a log *P* between 2.7 and 3.1, the apparent log *P* for the *stratum corneum* intercellular lipids.

A further study by Zhang et al. (2013) probed further the relationship between solvent/vehicle effects and flux. They examined flux, solubility of permeants in the *stratum corneum* and the permeability coefficient, k_p, for the data set of phenolic compounds of similar size used in their previous studies. In this case, they examined the effects of widely used, highly lipophilic vehicles—mineral oil (MO) and isopropyl myristate (IPM)—on skin transport; the former is a widely used ingredient in skin moisturising products, whereas the latter has shown an ability as an enhancer of transdermal absorption. Diffusion, spectroscopy and microscopy experiments were carried out (as in Zhang et al. 2011), and results were compared with the solvent systems reported in their earlier study. They found that maximum flux was similar for both solvent systems but that fluxes from IPM were higher for the more polar members of their data set, which was attributed to a higher rate of diffusivity. Very significantly they found that, while maximum flux for their data set was related directly to solubility in the *stratum corneum* and was independent of the solvent/vehicle, trends in the permeability coefficient were strikingly different. Specifically, an increase in penetrant lipophilicity increased the permeability coefficient for aqueous solvents and decreased the permeability coefficient for lipophilic solvents. Thus, overall, Zhang et al. concluded that the maximum flux for phenols with a similar molecular size and different lipophilicities was similar from mineral oil and water and higher for IPM and propylene glycol/water cosolvent systems. Insights from spectroscopy, microscopy and differential scanning calorimetry studies suggested that IPM increases lipid fluidity in the *stratum corneum*, increasing diffusivity and therefore flux for all phenols examined in these studies but particularly for the more polar phenols as the greater *stratum corneum* solubility of the more lipophilic phenols is balanced by their decreasing diffusivity.

Thus, this chapter provides a snapshot—and by no means an exhaustive review—of models that sit apart from the perceived mainstream approach of mathematical algorithms which dominate this field. This chapter therefore contains fewer algorithms describing percutaneous absorption than earlier chapters but, in significant contrast, offers significant mechanistic insight in a "bottom-up" approach. The studies from Roberts' group (Magnusson et al. 2004; Zhang et al. 2009, 2011, 2013) are highly significant in that they emphasise the importance of flux, rather than permeability, in their outputs. In doing so, they emphasise the importance of the former parameter, which is of greater clinical and toxicological significance than the permeability coefficient. Further, they have designed studies which have allowed substantial mechanistic insights to be proposed, particularly in the selection of their data set, in terms of each member having similar molecular weights and different lipophilicities and hydrogen bonding properties. This is a similar outcome to Mitragotri's (2003) study, but both achieve detailed mechanistic insights in very different ways. Thus, while it might be commented that the studies discussed in this chapter move more from quantitative to qualitative, they provide a significant level of insight perhaps lacking in more statistically based "top-down" models. For example, Magnusson et al. (2004) describes the issues of data consistency and its implication to the development of precise mathematical models, which have been elegantly elaborated upon in Zhang's studies. In addition, these studies do also raise the significant issue of how formulation is discussed in model development and, in that context, particularly in the light of the study by Gregoire et al. (2009), how approximations and, potentially, indirect measurements of formulation-associated phenomena may limit model quality and applicability. This subject is discussed in greater detail in Chap. 8.

References

Bast GE (1997) Influence of solubility and permeant size on absorption and metabolism of xenobiotics in rabbit skin. Hum Exp Tox 16:435–440

Borras-Blasco J, Diez-Sales O, Lopez A, Herraez-Dominguez M (2004) A mathematical approach to predicting the percutaneous absorption enhancing effect of sodium lauryl sulphate. Int J Pharm 269:121–129

Borras-Blasco J, Lopez A, Morant MJ, Diez-Sales O, Herraez-Dominguez M (1997) Influence of sodium lauryl sulphate on the in vitro percutaneous absorption of compounds with different lipophilicity. Eur J Pharm Sci 5:15–22

Cornwell PA, Barry BW (1993) The routes of penetration of ions and 5-fluorouracil across human skin and the mechanisms of action of terpene skin penetration enhancers. Int J Pharm 94:189–194

Cronin MTD, Dearden JC, Gupta R, Moss GP (1998) An investigation of the mechanism of flux across polydimethylsiloxane membranes by the use of quantitative structure-permeability relationships. J Pharm Pharmacol 50:143–152

Dal Pozzo AD, Donzelli G, Liggeri E, Rodriguez L (1991) Percutaneous absorption of nicotinic acid derivatives in vitro. J Pharm Sci 80:54–57

Deen W (1987) Hindered transport of large molecules in liquid-filled pores. AIChE 33:1409–1425

Diez-Sales O, Perez-Sayas E, Martin-Villodre A, Herraez-Dominguez M (1993) The prediction of percutaneous absorption I. Influence of the dermis on in vitro permeation models. Int J Pharm 100:1–7

Diez-Sales O, Watkinson AC, Herraez-Dominguez M, Javaloyes C, Hadgraft J (1996) A mechanistic investigation of the in vitro human skin permeation enhancing effect of Azone®. Int J Pharm 129:33–40

Edwards D, Langer R (1994) A linear theory of transdermal transport phenomena. J Pharm Sci 83:1315–1334

Flynn GL (1990) Physicochemical determinants of skin absorption. In: Gerrity TR, Henry CJ (eds) Principles of route-to-route extrapolation for risk assessment. Elsevier, New York, pp 93–127

Gregoire S, Ribaud C, Benech F, Meunier JR, Garrigues-Mazert Guy RH (2009) Prediction of chemical absorption into and through the skin from cosmetic and dermatological formulations. Br J Dermatol 160:80–91

Hatanaka T, Inuma M, Sugibayashi K, Morimoto Y (1990) Prediction of skin permeability of drugs. I. Comparison with artificial membrane. Chem Pharm Bull (Tokyo) 38:3452–3459

Higuchi T, Davis SS (1970) Thermodynamic analysis of structure-activity relationships of drugs: prediction of optimal structure. J Pharm Sci 59:1376–1383

Johnson ME, Blankstein D, Langer R (1995) Permeation of steroids through human skin. J Pharm Sci 84:1144–1146

Kasting G, Smith R, Cooper E (1987) Effect of lipid solubility and molecular size on percutaneous absorption. In: Shroot B, Schaefer HB (eds) Pharmacology and the skin, vol 1. Karger, Basel, pp 138–153

Kim YH, Ghanem AH, Higuchi WI (1992) Model studies of epidermal permeability. Semin Dermatol 11:145–156

Lai P, Roberts MS (1999) An analysis of solute structure human epidermal transport relationships in epidermal iontophoresis using the ionic mobility: pore model. J Cont Rel 58:323–333

Lai P, Roberts MS (1998) Epidermal iontophoresis: II. Application of the ionic mobility-pore model to the transport of local anesthetics. Pharm Res 15:1579–1588

Le VH, Lippold BC (1995) Influence of physicochemical properties of homologous esters of nicotinic acid on skin permeability and maximum flux. Int J Pharm 124:285–292

Magnusson BM, Anissimov YG, Cross SE, Roberts MS (2004) Molecular size as the main determinant of solute maximum flux across the skin. J Invest Dermatol 122:993–999

McCafferty DF, Woolfson AD, Moss GP (2000) Novel bioadhesive delivery system for percutaneous local anaesthesia. Br J Anaesth 84:456–458

McCafferty DF, Woolfson AD (1993) New patch delivery system for percutaneous local anaesthesia. Br J Anaesth 71:370–374
Menon GK, Elias PM (1997) Morphologic basis for a pore-pathway in mammalian stratum corneum. Skin Pharmacol 10:235–246
Mitragotri S, Johnson ME, Bankschte D, Langer R (1999) A theoretical analysis of partitioning, diffusion and permeation across lipid bilayers. Biophys J 77:1268–1283
Mitragotri S (2003) Modelling skin permeability to hydrophilic and hydrophobic solutes based on four permeation pathways. J Cont Rel 86:69–92
Morimoto Y, Hatanaka T, Sugibayashi K, Omiya H (1992) Prediction of skin permeability of drugs: comparison of human and hairless rat skin. J Pharm Pharmacol 44:634–639
Peck KD, Ghanem AH, Higuchi WI (1994) Hindered diffusion of polar molecules through, and effective pore radii estimates of, intact and ethanol treated human epidermal membrane. Pharm Res 11:1306–1314
Roberts M, Cross S, Pellett M (2002) Skin transport. In: Walters KA (ed) Dermatological and transdermal formulations. Marcel Dekker, New York, pp 89–195
Roberts WJ, Sloan KB (2000) Prediction of transdermal flux of prodrugs of 5-fluorouracil, theophylline, and 6-mercaptopurine with a series/parallel model. J Pharm Sci 89:1415–1431
Selzer D, Abdel-Mottaleb MMA, Hahn T, Schaefer UF, Neumann D (2013) Finite and infinite dosing: difficulties in measurements, evaluations and predictions. Adv Drug Del Rev 65:278–294
Singh P, Roberts MS (1996) Local deep tissue penetration of compounds after dermal application: structure–tissue penetration relationships. J Pharmacol Exp Ther 279:908–917
van der Bondi A (1964) Waals volumes and radii. J Phys Chem 68:441–451
Williams AC, Barry BW (1991) Terpenes and the lipid-protein partitioning theory of the skin penetration enhancement. Pharm Res 8:17–24
Zhang Q, Grice JE, Li P, Jepps OG, Wang G-J, Roberts MS (2009) Skin solubility determines maximum transepidermal flux for similar size molecules. Pharm Res 26:1974–1985
Zhang Q, Li P, Liu D, Roberts MS (2013) Effect of vehicles on the maximum transepidermal flux of similar size phenolic compounds. Pharm Res 30:32–40
Zhang Q, Li P, Roberts MS (2011) Maximum transepidermal flux for similar size phenolic compounds is enhanced by solvent uptake into the skin. J Cont Rel 154:50–57

Chapter 7
Squiggly Lines and Random Dots—You Can Fit Anything with a Nonlinear Model

Introduction

The application of nonlinear methods to the field of predictive algorithms in percutaneous absorption is not large, as many models produced have been linear in nature. Often, in the physical sciences, nonlinear relationships receive little attention and are seemingly given little credence compared to linear models. This is ironic, particularly in the context of the large amount of research into pharmacokinetic models of skin absorption, such as those reported previously (Kubota and Twizell 1992; Kubota et al. 1993), which often consider processes other than absorption. These, and similar, studies are based on the work of Chandrassekaran et al. (1976) which considered the binding of skin permeants to specific structural components within skin—and their temporary localisation and immobilisation within the skin—in the context of a Langmuir isotherm. Their results indicated concentration-dependent changes in lag and relaxation times after drug removal/absorption.

Kubota et al. employed a "random walk" approach to model the percutaneous absorption of timolol from a transdermal patch device. Their model, which comprised a one-dimensional homogenous membrane, did not require the level of detail normally associated with random walk methods to account for heterogeneities in outputs, and resulted in algorithms identical to finite difference schemes (Kubota et al. 1991). More complex models were also considered, employing additional skin layers to explore absorption kinetics. For example, a nonlinear "dual sorption" model was considered to evaluate the percutaneous absorption of timolol (Kubota and Twizell 1992; Kubota et al. 1993). Other researchers, such as George et al. (2004) and George (2005), have used similar approaches and confirmed the findings of earlier studies. It does remain, however, that the majority of models, particularly algorithms which consider skin absorption/uptake, consider the permeation only and do not consider subsequent processes, such as clearance, in any significant manner. The reader is directed to the excellent contribution of Roberts et al. (1999) for a more

G.P. Moss et al., *Predictive Methods in Percutaneous Absorption*,
DOI 10.1007/978-3-662-47371-9_7

comprehensive discussion of this particular subject. In addition, a number of nonlinear systems are associated with finite-dose scenarios, including patch devices and systems containing volatile solvents, and are discussed in Chap. 8.

While some algorithms for percutaneous absorption such as Lien and Gao's model (Lien and Gao 1995) described in Chap. 4 (Eq. 4.1) and which contains a quadratic function have been described earlier, their appearance in the literature has been infrequent and their further application minimal, possibly reflecting the perceived utility of such models due to their nonlinearity. Roberts et al. (2002) suggested that nonlinearity can occur in a system due to interactions between the solute and either its vehicle or the skin, but that such situations are unlikely at low concentrations. This implies that a degree of control can be exerted over the system to ensure the maintenance of desired properties and that if nonlinearity arises, it may be considered an artefact of the system rather than a genuine experimental condition. This comment also sits in the context of the use of infinite doses in most systems from which the models discussed thus far have been discussed (see Chap. 2).

The aim of this chapter is to examine the use of methods of analysis based on nonlinear principles which are generally applied to infinite-dose systems, such as those employed in the development of most models since Flynn (1990) and Potts and Guy (1992).

Application of a Nonlinear Multiple Regression Model to Skin Permeability

A significant use of nonlinear multiple regression was presented by Wilschut et al. (1995), who analysed the reliability of five models of skin permeability. They used a database of 123 measured permeability coefficients for 99 different chemicals and considered relevant physicochemical properties as well as experimental conditions. In collating their database, they were very precise in defining appropriate criteria for inclusion, which were that:

1. a permeation coefficient was presented as a result of the survey or could be derived from it
2. only chemicals applied to the skin dissolved in water were considered
3. the experimental conditions were fully known; this meant that data in secondary sources, such as reviews, were not included unless the primary source was also available
4. the permeation was measured under similar experimental protocols and conditions
5. the physicochemical descriptors investigated—log P and molecular weight (MW)—were known for all compounds in the data set.

In the context of future studies, point (5) might be considered an issue, in that the Wilschut method makes the assumption that only log P and MW were significant to the skin permeability process. Indeed, it should be reflected upon that other models were considering other descriptors and considering them to be of significance to skin permeation. These models are considered in earlier chapters (Chaps. 3 and 4).

They estimated the coefficients of five different skin permeability models, previously published in the literature (Brown and Rossi 1989; Fiserova-Bergerova et al. 1990; McKone and Howd 1992; Guy and Potts 1993; Robinson 1993). Multiple nonlinear least squares regression analysis was applied to estimate the regression coefficients of the five models. Of the five models, only the Guy and Potts (1993) model was fitted by multiple linear regression analysis. The output for each analysis consisted of the regression coefficients, the results of the Student's *t* test and the residual variance ($\log_{10}(k_{\text{p-observed}}/k_{\text{p-expected}})$. For purposes of validation, the data set was randomly split into training and validation sets, which also considered equal distribution of the physicochemical descriptors in the data set. The training set was used to estimate the regression coefficients of the model, while the validation set was used to estimate variance and to statistically test the model (i.e. in terms of residual variance, F test and to compare the regression coefficients of the model).

Interestingly, they included a breakdown of the experimental conditions in their study. This highlighted that three types of diffusion cell were used (including two permeability data measured in flow-through diffusion cells); that for 29 experiments, the anatomical site from which the skin was harvested was unknown; and that the mean temperature for 115 in vitro experiments considered was 30 °C. In 15 % of experiments, the composition of the receptor phase was unknown, and where it was stated, water was the overwhelming choice. Other features, such as occlusion, whether a study was conducted in vivo (which was removed to aid consistency in the final data set) or in vitro, the chemical method of analysis and thickness and type of skin, were also recorded. It is interesting that the authors were very detailed not only in describing the different experimental conditions but also in commenting that such potential sources of variance are not considered in their model. Indeed, the authors contended that such sources *cannot* be considered by their model as it would result in far too few permeability values being considered and produce a model of little relevance—such an inference can clearly be extended to other studies and may inform considerations of model fit and statistical quality. It is also interesting to consider these comments both in their absence from other studies and in the context of studies which examine subsets (see Chap. 4) of larger data sets. Often, those studies report highly significant models, but the inference from Wilschut's study is that such highly correlated models may, due to the small data sets used, have little relevance or scope in a global sense and may only be relevant to a small number of potential skin permeants. Such wide variation in experimental conditions from which the permeability data were derived may influence substantially the residual variance reported between observed and estimated permeation coefficients. Wilschut et al. further stated that, ideally, consistent experimental protocols would be needed to improve model quality. They commented that the data to do that were not yet available. Almost 20 years later, it is reasonable to comment that that situation has changed little as the data that have appeared in the literature (and the protocols used to generate these data) since 1995 reflect the needs of the originators rather than those abstracting such data to develop permeability models.

Wilschut et al. found that three models—those based on studies published previously by McKone and Howd (1992), Guy and Potts (1993) and Robinson (1993)—were considered to provide reliable estimations of the skin permeability coefficient. While they commented that the model of Guy and Potts was relatively simple as it considered only the *stratum corneum* as a lipid barrier, they stated that the McKone and Howd model and the Robinson model were more complex as they considered permeability through a watery layer on the skin and beneath the skin, respectively. They also commented that the McKone and Howd and the Robinson models consider permeation via a polar pathway through the *stratum corneum*; diffusion through this pathway and through the aqueous layers of underlying viable epidermis is separately modelled, and this was regarded by Wilschut et al. as more realistic than the approach taken by Guy and Potts (1993). They also found that the models reported by McKone and Howd were more able to accurately predict the skin permeation of highly hydrophilic and highly lipophilic chemicals compared to the model by Guy and Potts. They did, however, comment that the Guy and Potts model was relatively good at predicting the middle of the range of lipophilicities modelled—that is, $-1 < \log_{10}K_{ow} < 5$ and that it was less accurate at lower and higher lipophilicities.

Based on their statistical analysis, they concluded that the best model was a modified version of the Robinson (1993) model. Their analysis resulted in the modification of the models by Guy and Potts and Robinson to include $MW^{0.5}$ as an independent parameter as it gave a better fit than MW in the original models:

$$k_p(\text{cm/h}) = \frac{1}{\frac{1}{K_{psc}+K_{pol}} + \frac{1}{K_{aq}}} \tag{7.1}$$

where

$$\log K_{psc} = -1.326 + 0.6097 \log K_{ow} - 0.1786\,\text{MW}^{0.5} \tag{7.2}$$

is the permeation coefficient of the lipid fraction of the *stratum corneum*,

$$K_{pol} = \frac{0.0001519}{\sqrt{\text{MW}}} \tag{7.3}$$

is the permeation coefficient of the protein fraction of the *stratum corneum*, and

$$K_{aq} = \frac{2.5}{\sqrt{\text{MW}}} \tag{7.4}$$

where K_{ow} is the octanol–water partition coefficient and MW is the molecular weight.

When Wilschut compared the original Robinson model with their iteration, they found that the original model underestimated skin permeation, and that the new model altered the influence of MW in the final model, increasing the significance of diffusion through the protein fraction of the *stratum corneum*.

Thus, they concluded that it was possible to make an optimum choice for a skin permeation model in connection with a specific data set (another underlying point not widely considered, but addressed albeit obliquely in "subset" studies). Their revised version of the Robinson (1993) model has the best performance (in terms of having the smallest residual variance) *for the data set studied*. They also commented that MW was not correctly considered in any of the models—other than Robinson's—underlining the nonlinear nature of their analysis and of the skin permeability data set.

In more recent years, the use of nonlinear models has tended to focus on the use of Machine Learning methods, such as fuzzy logic, neural networks and Gaussian processes (GPs). The remainder of this chapter will focus on those methods, and in particular at the reasons why they appear to offer better models; why they are often criticised as being of little relevance to the real world; and why, after a only small number of publications, studies in specific areas tend to find little or no audience.

Fuzzy Logic and Neural Network Methods for the Prediction of Skin Permeability

As described in previous chapters, models relating skin permeability to physicochemical properties of potential penetrants have classically focused on findings drawn from experimental studies. These experiments are normally in vitro models, described in Chap. 2, which involve measuring the amount of chemicals that permeate into and across skin (usually human or a suitable alternative, such as porcine skin) over a set period of time (usually 24–72 h). The amount of drug absorbed over time is determined, and from this, the flux of permeation (usually the gradient of the zero order, steady-state part of the drug release profile) is calculated. The flux and its concentration-corrected counterpart, the permeability coefficient, are commonly used to describe the process of permeation in algorithms of skin permeability. This subject is discussed in detail in earlier chapters and is described in greater depth elsewhere (Moss et al. 2002; Williams 2003; Mitragotri et al. 2013). Thus, within this chapter, the principles described in these texts are discussed not only in the context of key studies by Flynn (1990) and Potts and Guy (1992), but also in the light of Wilschut's findings (Wilschut et al. 1995).

In general, "Machine Learning" methods were defined in 1959 by Arthur Samuel as being within a "field of study that gives computers the ability to learn without being explicitly programmed". These methods, as applied to percutaneous absorption, are generally considered to be *supervised learning* methods. This is where the computer is given inputs and outputs and aims to map the former to the latter. While this encompasses most of the Machine Learning methods applied to the field of percutaneous absorption, other methods, including classification-based approaches, have also been considered; some of these methods may be categorised as *unsupervised learning* methods, which are more commonly used in pattern recognition studies of higher dimensional data.

One of the earliest such methods applied to the prediction of skin permeability was "fuzzy" logic. For example, Pannier et al. (2003) used the adaptive neural fuzzy interference system to model skin permeability.

Like most modelling methods, fuzzy logic essentially maps inputs to outputs. For percutaneous absorption, the output is usually the skin permeability coefficient (or perhaps the flux) and the inputs are significant physicochemical descriptors of a molecule, or a data set of molecules; commonly used descriptors include measures of lipophilicity, such as log P or log K_{ow}, MW or molecular volume, melting point and hydrogen bond activity (i.e. the count of hydrogen bond acceptor and donor groups on a molecule). The difference in the fuzzy model is the method used to map the input to the output; independent of the methods used, all traditional modelling methods impose a mapping based on known information and a set of conventions, or rules, are used to develop the model. Such rules may include the assumed nature of the output, i.e. a linear model. An alternative to this is to use a model free from such restrictions which impose no rules on the system. In such cases, the rules are developed through the use of clustering algorithms which divide the data into natural groups, after which mapping of inputs to outputs is optimised. The rules can be either imposed by the researcher developing the method or determined from the data. It can be "crisp" (i.e. true or false statements) or "fuzzy", where the "crispness" of the result is modified based on the nature of the data; if it lies on a continuum, it may help particular studies to avoid arbitrary cut-off [i.e. MW greater than, or less than, 150 Da, as in Flynn (1990)]. Thus, if the data have been clustered into groups where membership of each group was either partial or by degrees of belonging, as opposed to a specific "yes" or "no" to membership, then such an arrangement would be considered "fuzzy".

Thus, Pannier et al. (2003) developed three models of skin permeability using a subtractive clustering technique, which defined structures within the data and allowed rules governing permeability to be defined. The models developed were able to predict skin permeability as well as, or better than, previously published algorithms with fewer inputs—correlation coefficients, as r^2, for the three "fuzzy models" of Flynn, Potts and Guy and Abrahams, were 0.828, 0.973 and 0.959, respectively. The models developed were related to log K_{ow} and MW (the "Flynn fuzzy model"; n = 94), and to $\sum \alpha_2^H$ and log K_{ow} (the "Potts and Guy fuzzy model"; n = 37. The third model, the "Abraham fuzzy model", was a variation on the Potts and Guy fuzzy model where the data set was slightly larger (n = 53) and MW was replaced by molecular volume. The authors commented that, by testing combinations of inputs, they could determine the best fuzzy model and also discern those descriptors most important to the process of skin permeability. Further, they demonstrated improvements over the traditional modelling methods and commented that further improvements in clustering methods and the range of selected inputs could improve further model quality.

Similarly, Keshwani et al. (2005) applied fuzzy logic—in this case, a rule-based Takagi-Sugeno method—to a skin permeability data set. It is interesting to note, in the context of methodological developments and the modelling of small subsets, that the authors justified the use of this method due to the "sparseness and ambiguity of available data". They analysed a large data set (n = 140) and used lipophilicity, MW and experimental temperature (which was a combination of skin surface temperatures and water bath temperatures from a diverse range of experiments) as inputs. In comparison with simple regression methods [by comparison of r^2 and root-mean-squared error (RMSE)], they found that their fuzzy model was superior, when compared with the same inputs.

It is important to note that, despite the obvious improvement in model quality and the success of such models, they have found little or no widespread use in the field of percutaneous absorption. Indeed, it is common that a small number of studies which use such methods are published which provide improved models but which may be outside the scope of dermal absorption scientists to fully apply to this field. This may be due to the lack of a defined output (an algorithm) or the technical aspects of model development [i.e. access to specific software, such as MATLAB, and to additional codes often used within such packages, as described by Pannier et al. (2003)], or to the often expensive requirement for expensive software packages.

Another successful field of sporadic interest to the modelling of percutaneous absorption is the application of artificial neural networks (ANNs). ANNs are biologically inspired computer programs which aim to mimic the perceived way in which the human brain processes information. They detect patterns and relationships within a data set and "learn", or are trained systematically through experiential modifications, rather than from specific programming and rule development or application. ANNs are formed from numerous, often hundreds, of single processing elements (PEs) which are connected via a series of coefficients, or weightings, each of which signifies the relative importance of connections within the network (Fig. 7.1).

The inputs of each PE within a specific network are specifically weighted. They also have specific transfer, or transformation, functions and a single output (generally, in skin absorption models, this would be a prediction of the permeability coefficient). Data may feed backwards, or forwards, into different functions of the network, influencing the nature of the output (Fig. 7.2).

The use of transformation functions may introduce nonlinearity into the resultant model, but such phenomena are optimised for each PE within a network on order to reduce errors in predictions. Once such functions have been optimised and validated (with test and training data set, or subsets of a larger data set), then they can be used to provide predictions of skin permeability for new chemicals which are not in the original data set but which sit within its molecular space (Agatonovic-Kustric and Beresford 2000).

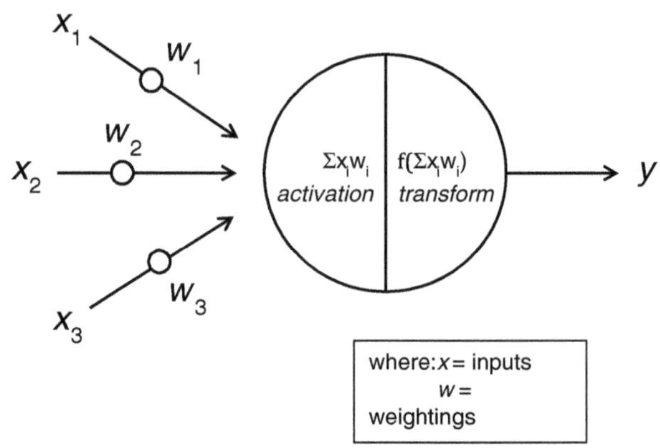

Fig. 7.1 Model of an artificial neuron. From Ashrafi et al. 2015; modified from Agatonovic-Kustrin and Beresford (2000)

These methods have been widely employed in the pharmaceutical sciences; not only in modelling skin absorption but more broadly in, for example, formulation studies as an alternative to response surface methods (Agatonovic-Kustric et al. 1999), in assessing permeation across a polydimethylsiloxane membrane (Agatonovic-Kustric et al. 2001; see Chap. 5), optimisation of solid dosage form design (Bourquin et al. 1997, 1998; Takahara et al. 1998) and emulsion formulation (Alany et al. 1999; Fan et al. 2004), gene classification and protein structure prediction and sequence classification (Sun et al. 1997; Wu 1997; Milik et al. 1995). Recent research has also seen the application of genetic algorithms to pharmaceutical problem domains, specifically in the context of quantitative predictive models of drug absorption using a QSPR-based approach as described in previous chapters, where the predicted permeability across a biological membrane is related to key physicochemical descriptors of molecules in a data set (Willett 1995; So and Karplus 1996, 1997a, b). They have even been applied to clinical studies, such as the analysis of skin disease classified by Kia et al. (2013). Degim et al. (2003) applied a previously published partial charge equation and ANN methods to develop a skin permeability model. Using a data set taken from the literature (n = 40), an ANN was developed whose outputs correlated very well with experimental values (r^2 = 0.997), providing a precise model for estimating percutaneous absorption (Ashrafi et al. 2015).

Chen et al. (2007) used ANNs to predict the skin permeability coefficients of novel compounds. They used a large data set (n = 215) which was described by the descriptors reported previously by Abrahams et al. (1997). Their data were subdivided into various subsets, four of which were used to train and validate the chosen models (an ANN model and a simple multiple linear regression model, which was used to benchmark the ANN model) and the remainder was used to test the models. They reported that the ANN model was nonlinear in nature and was significantly better, in terms of its statistical and predictive performance, than the linear regression model. For example, the multiple linear regression model performance was weaker in its statistical and predictive performance

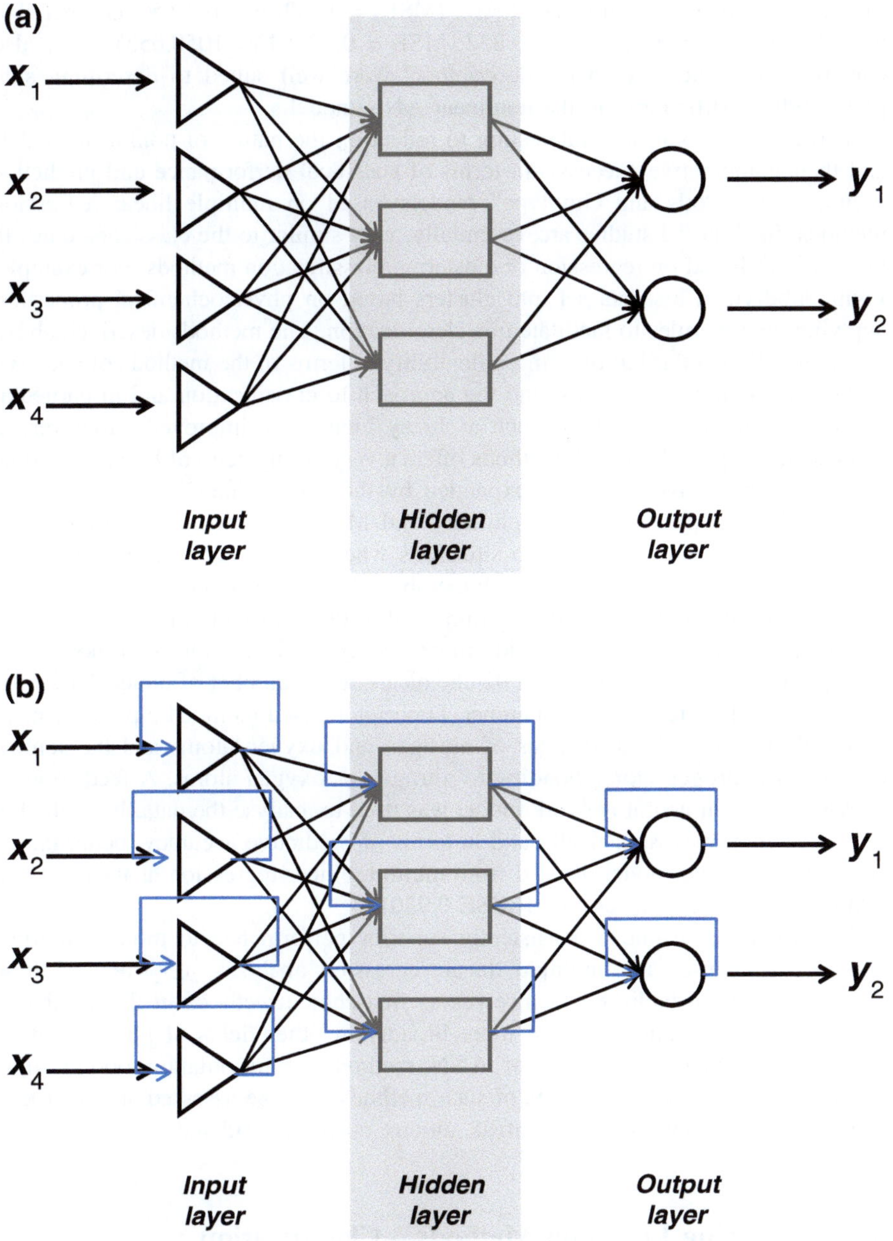

Fig. 7.2 Schematic model of a feed-forward network (**a**, *top*) and a feed-backward network (**b**, *bottom*). From Ashrafi et al. 2015; modified from Agatonovic-Kustrin and Beresford (2000)

($n = 215$; $r^2 = 0.699$; mean-squared error (MSE) = 0.243; $F = 493.556$) compared to the ANN model ($n = 215$; $r^2 = 0.832$; MSE = 0.136; $F = 1050.653$). They also concluded that the "Abrahams descriptors" were well suited to describing skin permeability, particularly in the nonlinear ANN model.

Thus, at this point, it is interesting to reflect on the nature of nonlinear models and their comparative success—in terms of statistical performance and predictive accuracy—to "Potts and Guy-type" models based on multiple linear regression methods. Such novel studies are, essentially, very similar to the classical studies in that they are based on regression or clustering/classification methods. For example, Flynn subdivided his data set into clusters based on physicochemical properties, applying distinct rules to facilitate this classification. The methods described above are essentially similar but offer more flexibility in terms of the methods of analysis, particularly nonlinear analysis, and the approach to classification and in particular boundaries, in which methods such as fuzzy logic have improved. However, in essence, the approach of such methods offers a very strong echo of Flynn's original approach. They have also been expanded by the use of "new" descriptors, progressing from 2 parameters (lipophilicity and MW) through the adoption of the so-called Abrahams descriptors to situations where, potentially, several thousand descriptors can be determined for each member of a data set and used in its analysis. An example of this is the study by Lim et al. (2002), in which molecular orbital parameters were employed alongside more widely used descriptors to model skin absorption. They used a data set of 92 chemicals, and a number of molecular orbital terms were calculated for each member. Descriptors used included dipole moment, polarizability, the sum of charges of nitrogen and oxygen atoms and the sum of charges of hydrogen atoms bonding to nitrogen or oxygen atoms. A feed-forward back-propagation neural network model was used to analyse the data. It resulted in a model which was, statistically and in terms of predictive accuracy, better than a conventional linear model derived from multiple linear regression analysis (ANN: RMSE 0.528; linear regression: RMSE 0.930).

Nevertheless, despite a consistently superior performance to more traditional approaches—particularly multiple linear regression analysis—very few of these techniques have established themselves as first-choice methods in the prediction percutaneous absorption or even more broadly in other fields of pharmaceutical development, such as the use of ANN methods in formulation optimisation. Therefore, the real-world benefits of such methods must be assessed and their lack of uptake by pharmaceutical scientists, among others, considered.

More Machine Learning Methods—Classification and Gaussian Process Models

In general, ANN and related Machine Learning methods require specific expertise in computer programming statistics which may be outside the reach of many physical scientists, which may impact on the ability to apply such specific and high-level applications from one field into another. In doing so, it echoes the comments by Cronin and Schultz (2003) regarding the need for specialist expertise in all aspects of model development and analysis. Indeed, this may be reflected in, for example, the work of Danick et al. (2013) in developing a spreadsheet-based model for estimating bioavailability of chemicals from dermal exposure. Implicit in such a study is the simple utility required to make a method work broadly in a different field. While some of the Machine Learning approaches suggest that Potts and Guy's model, and the general approach of multiple linear regression, is inferior to the use of any number of Machine Learning studies, they also suggest that ease of use, transparency and broad utility that do not require specialist (and often very expensive) software are significant advantages. So too is the use of descriptors which are readily interpreted and relevant to physical scientists and which are, again, relatively straightforward to determine and which do not require expensive software packages. It would therefore appear that, currently, the limitations in Machine Learning methods outweigh their advantages. It also sends a message to those who develop and use such specific software-based approaches, which is that their utility will improve significantly if they are made more accessible and more readily interpretable by potential users in other fields.

In an example of this approach, Baert et al. (2007) employed a classification Machine Learning method to analyse a data set of 116 compounds (mostly drugs). The authors calculated and compared several models. Their initial 9-parameter multiple linear regression model only explained 40 % of the variability. They used an expanded range of computed molecular descriptors and developed a predictive algorithm based on log k_p. They used a classification method—the classification and regression trees (CART) technique—which was validated by an additional twelve chemicals which were within the molecular space of their data set but not members of it. Following classification, the final model was determined by multiple linear regression analysis and resulted in a 23-term model. To avoid over-parameterisation and to simplify their model, they employed both the Kubinyi function and Akaike's information criterion.[1] Their analysis returned a q value of 9.45, well above the normal minimal value of 4 considered for the development of a linear model. Thus, they considered the inclusion of additional descriptors in their model but found that application of the Kubinyi function gave decreased values when more variables were added to the model, suggesting over-fitting. The latter test showed a biphasic asymptotic decrease, and their final model was a 10-parameter expression which the authors claimed addressed some of the concerns discussed above and presented a compromise between the statistical quality of the model, and

[1]The Kubinyi function is a statistical parameter closely related to the Fisher ratio (F). Whereas the Fisher ratio (F) is often sensitive to changes in small d values, and poorly sensitive to changes in large d values, the Kubinyi function avoids these issues. In general, a larger Kubinyi values suggest a better linear equation. The Akaike's information criterion is an indicator of the relative quality of a model for a particular set of data and thus is used in model selection.

its predictive ability as it modelled over 70 % of the variability, and its mechanistic complexity and transparency as the addition of further parameters to the model resulted only in marginal increases in its quality. Their proposed linear model is given as follows:

$$\begin{aligned}\log k_p(\mathrm{cm/s}) = &- 6.243\,(+2.12 \times 10^{-1}) \\ &- 3.14\,(+6.17 \times 10^{-2})\ \mathrm{H.050} \\ &- 1.03\,(+2.09 \times 10^{-1})\ \mathrm{Hypertens.50} \\ &+ 1.04 \times 10^{-1}(+5.73 \times 10^{-2})\ \mathrm{ALOGP} \\ &- 4.84 \times 10^{-4}(+1.05 \times 10^{-4})\ \mathrm{SRW09} \\ &+ 1.50 \times 10^{-1}(+3.09 \times 10^{-2})\ \mathrm{RDF075m} \\ &- 1.39 \times 10^{-1}(+2.99 \times 10^{-2})\ \mathrm{H.052} \\ &- 4.84 \times 10^{-1}(+8.65 \times 10^{-2})\ \mathrm{T.(S..F)} \\ &+ 4.77 \times 10^{-1}(+1.10 \times 10^{-1})\,\mathrm{C.025} \\ &- 10.60\,(+2.73)\ \mathrm{R1m+} \\ &- 6.15\,(+2.00)\ \mathrm{RTm+}\end{aligned} \tag{7.5}$$

where

H.050 (atom-centred fragment) represents the number of hydrogen atoms attached to a heteroatom
Hypertens.50 (molecular property class) is the Ghose-Viswanadhan-Wendoloski 50 % antihypertensive druglike index
SRW09 is the self-returning walk count of order 09
RDF075 m is the radial distribution function 7.5, which is weighted by atomic masses (i.e. the corrected probability distribution associated with finding an atom in a spherical volume with radius r)
H.052 is the number of hydrogen atoms attached to C^0(sp3) with one halogen attached to the next C
T.(S..F) is the sum of topological distances between S and F atoms
C.025 is the atom-centred fragment R-CR-R
R1m+ and RTm+ are, respectively, GETAWAY class descriptors describing the maximal autocorrelation of lag 1 and the maximal index, both of which are weighted by atomic masses.

This model also had the lowest room MSE of prediction, 0.73, of the models evaluated, while the CART model had the worst (1.76). Comparison of this regression model with other published studies indicated that it was comparable in terms of its statistical quality. Thus, Baert et al. classified their data set into a distinct number of permeability classes using the CART method in order to obtain a selected number of model penetrants; this output also indicated that the OECD reference compounds caffeine, benzoic acid and testosterone were classified into different clusters. Further, models of good statistical quality were obtained using parameters that related to the lipophilic nature of penetrants and to descriptors of 3D- and 2D-molecular stereochemical complexity, and explained the skin permeability better than other descriptors. The use of the CART-clustering method indicated that, as penetrants became more lipophilic, the extra-dimensional information encoded in a three-dimensional molecular representation became less significant, while the opposite was found to be true for increasingly hydrophilic compounds.

Thus, there are several interesting outcomes from Baert's comprehensive and excellent study. Their analysis involve the use of a wide range of descriptors, effectively employed classification/clustering techniques and expressed—and dealt with—specific concerns of over-fitting when using a wide range of descriptors; this latter point is of huge significance in the acceptance and use of nonlinear or Machine Learning methods as the general perception is that such methods will automatically over-fit data, often therefore leading to nonlinear outcomes. It also interesting to therefore note that their approach used *linear* regression methods to relate log k_p to the significant molecular descriptors. However, the model still lacks accessibility, given the parameters returned as significant, and their utility in the field by non-experts in modelling. Thus, their approach has sadly found little further application within the field of percutaneous absorption.

More recently, GPs have found utility in a number of fields, and they were first applied to the problem domain of percutaneous absorption by Sun et al. (2008), who initially concluded that the patterns inherent in the data suggested a fundamental lack of linearity in the data.

The aim of the GP is to model the relationship between the inputs and the outputs. It begins with a set of N data items (x_n), such as permeability data (as k_p, or log k_p) which has corresponding output values, y_i. The GP model infers a function that relates the input descriptors to the output (i.e. relates the physicochemical descriptors of a molecule in the data set to its permeability coefficient) for the data set and then predicts skin permeability for a new compound. A range of Machine Learning methods were used by Moss et al. (2009), including simple linear regression, which is a linear regression method which uses iterated reweighted least squares training, and Gaussian process regression (GPR), which calculates the relationship between input and output via a nonlinear process. They used a large data set (n = 142) which was based on that published by Flynn (1990) and which was supplemented by data presented in the EDETOX database (available at www.ncl.ac.uk/edetox/index.html), data published elsewhere (i.e. Wilschut et al. 1995; Patel et al. 2002) and other additions which were described by Moss et al. (2006). Six descriptors were employed to describe their data [log P (predominately measured values taken from the literature, but, where no alternative was available, predicted values were used from the KOWWIN source), MW, the count of hydrogen bond donors and acceptors on a molecule, the solubility parameter (Fedors 1974) and the melting point].

The simple linear regression method is described by the expression:

$$y = y(x, w) = \sum_{i=1}^{d} w_i x_i + w_0 \tag{7.6}$$

where d is the dimensionality of the input space (i.e. the number of descriptors used to describe a molecule) and $w = (w_1; \ldots ;w_d; w_0)$, which is the weighted vector, where the weights are set so that the sum-squared error function is minimised on a training set.

Alternatively, the GPR model is a nonparametric method. In common with the other Machine Learning methods described above, it does not produce an explicit functional output (i.e. an algorithm), and it is assumed that the underlying function that produces the outputs, f(x), will remain unknown, but that the data are produced from a (infinite) set of functions with a Gaussian distribution in the function space. The Gaussian function is fully characterised by its mean and covariance. The mean is usually considered to be the "zero everywhere" function, and the covariance, $k(x_i, x_j)$, expresses the expected correlation between the values of f(x) at the two points x_i and x_j and in doing so defines "nearness" or similarity between data points within the data set and predictions made by the model. The GP model has a Gaussian distribution and its mean is defined as:

$$\mathrm{E}[y_*] = k_*^T \left(K + \sigma_n^2 I\right)^{-1} y \tag{7.7}$$

where k_* is the vector of covariance between the test point and the training data, K is the covariance matrix of the training data, σ_n^2 is the variance of an independent identically distributed Gaussian noise (which means that observations are noisy), K_*^T is the transpose of K_*, I is the identity matrix, and y is the vector of training targets.

The variance of the GP model is given by:

$$\mathrm{var}[y_*] = k(x_*, x_*) - k_*^T \left(K + \sigma_n^s I\right)^{-1} k_* \tag{7.8}$$

where $k(x_*, x_*)$ denotes the variance of y_*.

Moss et al. (2009), and in future studies described below, used the mean as the prediction and the variance as the error bars on the prediction. They used a number of performance measures, common the Machine Learning studies, to characterise the quality of their models. These included the normalised MSE (where the MSE is normalised by the variance of target values), improvement over the naïve model (ION), which indicates the degree of improvement of the model over the Naïve predictor:

$$\mathrm{ION} = \frac{\mathrm{MSE}_{\mathrm{naive}} - \mathrm{MSE}}{\mathrm{MSE}_{\mathrm{naive}}} \times 100\,\% \tag{7.9}$$

where MSE is the mean-squared error and $\mathrm{MSE}_{\mathrm{naive}}$ is the MSE of a naïve model (which is the arithmetic mean of experimental k_p values).

They also used the average negative log estimated predictive density (NLL) to describe their model's quality:

$$\mathrm{NLL} = \frac{1}{n}\sum_{n=1}^{N_{\mathrm{tst}}} -\log p(y_n|x_n) \tag{7.10}$$

where $-\log p(y_n|x_n) = \frac{1}{2}\log\left(2\pi\sigma_*^2\right) + \frac{(y_n - E[f_*])^2}{2\sigma_*^2}$, in which case σ_*^2 is the predictive variance.

Finally, the correlation coefficient, as r^2, was also used to characterise their model. In addition, principal component analysis (PCA, shown in Fig. 7.3) and data visualisation techniques (Fig. 7.4) were also employed to characterise the distribution of the data set membership. PCA indicated that the first two principal components accounted for two-thirds of the total variance, and also indicated that there is no linear relationship between log k_p and any of the six physicochemical descriptors examined. The authors suggested therefore that more complex nonlinear structures were apparent in their data set. This is re-enforced by Fig. 7.4, which compares all the relationships across the data set and which again suggests a significant nonlinear distribution of the data.

Analysis of their data set by Machine Learning (GP) methods, and comparison with QSPR methods, including the Potts and Guy (1992) algorithm, indicated that the GP method was vastly superior. The GP methods saw improvements in performance compared to the naïve model, whereas QSPRs performed poorly, producing worse predictions than the naïve model (−35.55 % in the case of the Potts and Guy algorithm), which is the mean of the data set, and a poor correlation (0.36). The single linear network model saw improvements over the QSPR model in ION (−35.55 (QSPR) vs. 11.20 (ION, 2 parameters—log P and MW) and 11.70 (ION, all six parameters), NMSE (1.48 vs. 1.02 and 1.00), and correlation coefficient (0.36 vs.

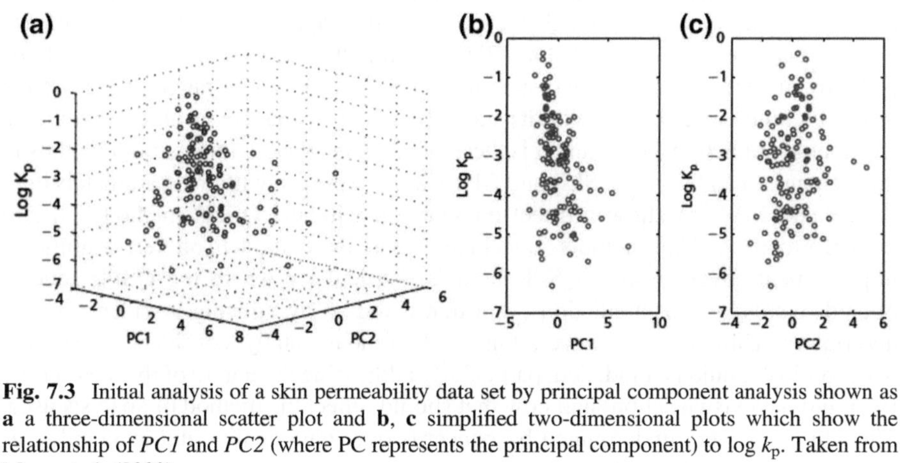

Fig. 7.3 Initial analysis of a skin permeability data set by principal component analysis shown as **a** a three-dimensional scatter plot and **b**, **c** simplified two-dimensional plots which show the relationship of *PC1* and *PC2* (where PC represents the principal component) to log k_p. Taken from Moss et al. (2009)

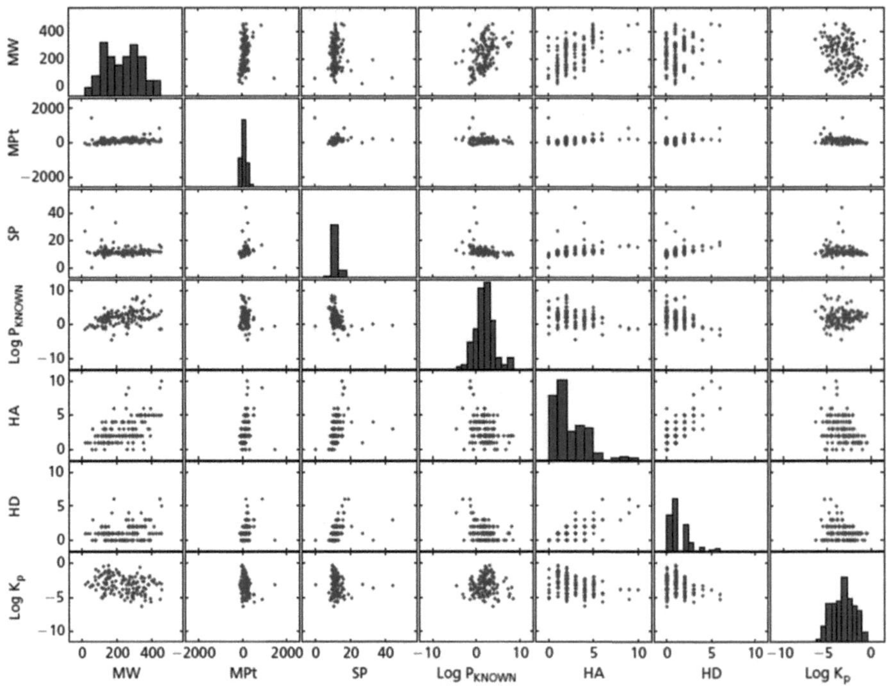

Fig. 7.4 A scatter plot matrix of a skin permeability data set. The diagonal (from *top left* to *bottom right*) shows the distribution of each molecular descriptor examined in this study. Taken from Moss et al. (2009)

0.38 and 0.43). The GP model saw further improvements compared to the SLR model, with lower NMSE (0.84 and 0.72), better ION (25.48 and 35.51) and improved r^2 (0.53 and 0.59) where figures relate to the 2- and 6-parameter GP models, respectively. This study does also indicate that additional descriptors do improve model quality, although the range of descriptors used is not as comprehensive as those used by Baert et al. (2007). While this data set did not use the Potts and Guy (1992) data set, its data are entirely within the stated range of that QSPR model and therefore should, with its stated bounds of variance, be able to provide a reasonable predictive model and it is therefore reasonable to compare both models in this manner. In addition, the SLR and GP methods were run with two parameters only (log P and MW) to allow a direct comparison with the QSPR method and still outperformed it. The authors concluded that their model offered significant improvements over existing QSPR models, particularly where the QSPR models offered relatively poor predictions, and that it did so by utilising a wider range of descriptors. Ultimately, however, this study fundamentally considered *how* skin permeability should be modelled, particularly addressing the nature of the data set (by PCA and data visualisation) and how that should reflect the method of analysis used. Their findings were underpinned by concerns over data quality and availability as discussed extensively in Chap. 9.

The key criticisms of the above study, in common with most Machine Learning approaches, are the lack of transparency (i.e. the absence of a readily interpretable algorithm relating inputs to outputs) and the adoption of nonlinear methods, which are generally perceived to be prone to over-fitting. Thus, GP and similar Machine Learning models fail to meet the key criteria of an algorithm for skin permeability—the ability to provide accurate predictions of permeability and mechanistic insight to the permeation process. While model performance in general and Baert's methods more specifically address the latter point, the use of simple methods of data visualisation to understand the nature of the data, and any resultant model, are key elements in beginning to address the issue of transparency.

Lam et al. (2010) attempted to address this issue by applying feature selection methods in order to determine the most significant descriptors in skin permeability for the data set described by Moss et al. (2009). This technique, in the context of skin permeability, produces a series of test sets which contain every possible permutation of the physicochemical descriptors used in the analysis. This allows the best model, in terms of its statistical performance, to be determined but it also ensures that the most significant combination of descriptors will be known, largely addressing the issue of model transparency. Their results again echoed their earlier study (Moss et al. 2009) which indicated that the data set examined (n = 142) was inherently nonlinear with respect to log k_p and that GP models had a better performance than both simple linear regression Machine Learning methods and QSPR models (those published by Potts and Guy (1992), Barratt (1995), Moss and Cronin (2002), which often performed significantly worse than the naïve model. They presented a wide range of GP models each of which used different covariance functions but found that all had similar performance, with the GPR and GPRARD methods, described above, generally exhibiting the best overall performance. In terms of feature selection, they found that the best models were those which used log P, the count of hydrogen bond donors and either MW or melting point to develop a GP model. The performance of this model is compared to the Potts and Guy (1992) predictions for a subsection of Flynn's data set in Fig. 7.5.

The significance of this study was that it indicated that descriptors used in modelling were, effectively, interchangeable to a degree. While this is perhaps not unexpected given the nature of the descriptors and their covariance (i.e. the relationship between log P, MW and melting point is well defined, usually for series of molecules such as *n*-alcohols), it is not something that is widely reported in this field, ostensibly because the output from previous models is mostly in the form of discrete algorithms which feature statistically significant parameters. It is also interesting to note that Lam's GP models reflect the importance of hydrogen bonding for a large data set, echoing the earlier findings of Robert's seminal series of papers (Roberts et al. 1995, 1996; Pugh et al. 1996).

In studies by both Lam et al. (2010) and Sun et al. (2011), the use of length-scale analysis also refined the model and gave more specificity to descriptor over-fitting. The inverse of the length-scale determines the relevance between the input and the output, with a low length-scale suggesting a high dependence of one descriptor on another. The length-scale analysis in both studies suggested that the addition of

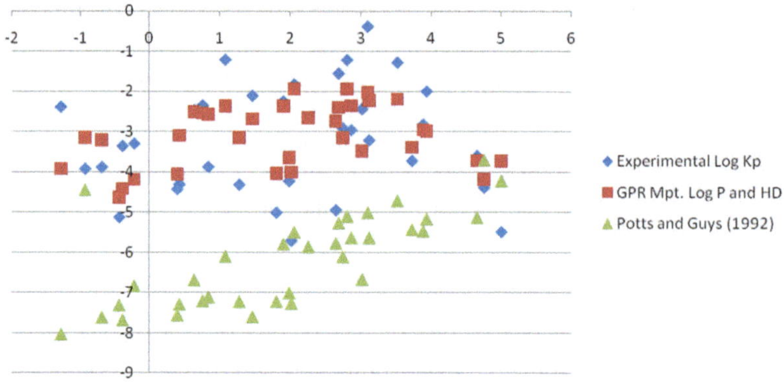

Fig. 7.5 A plot of log *K*p and log P for three data sets: *blue diamonds* are data from Flynn's original data set (Flynn 1990); *green triangles* are predictions generated from the Potts and Guy (1992) algorithm, and *red squares* are the predictions made with a 3-descriptor Gaussian process model (Lam et al. 2010). Modified from Lam et al. (2010)

further descriptors did not improve the statistical quality of the model, reflecting comments of excessive descriptor use by Baert et al. (2007). Indeed, the inclusion of the Fedor's solubility parameter was often seen to reduce the quality of GP models in which it was included, compared to models where it was not included. It should also be noted that, in a direct comparison with two-parameter QSPR models, Sun et al. (2011) demonstrated not only the superiority of the GP method but also the inherent nonlinearity of the absorption process, shown in Fig. 7.6 as a predictive "map" of skin permeability.

While length-scale analysis provides clarity and simplicity in Machine Learning model development, the apparent interchangeability of descriptors employed in an analysis of one of the largest data set (n = 142) of skin permeability may reflect the nature of previously published models which are significantly different but which use, in various forms, the same input data—or subsets of this data. The use of small data sets may bias the output of models to particular physicochemical parameters.

Further, Moss et al. (2011) expanded the GP method to consider a range of membranes (human, rat and mouse skin, and artificial (silicone) membranes) and also to consider the effect of data set size on model relevance and quality. Using their previously published methods, they developed a series of models (QSPRs and Machine Learning-based GP methods) for permeability in each data set (i.e. for each membrane). They then cross-validated the membranes, where possible, in order to establish if one membrane could act as a good predictor for the others, i.e. if silicone, mouse and rat permeability models had any relevance to human skin permeation.

As before, the GP models outperformed QSPR models for human, pig and rodent data sets. Poor correlations were found with the artificial membrane, which might suggest its lack of applicability as a replacement for skin, or it may reflect the size of the data set used in their study. Interestingly, in their use of different

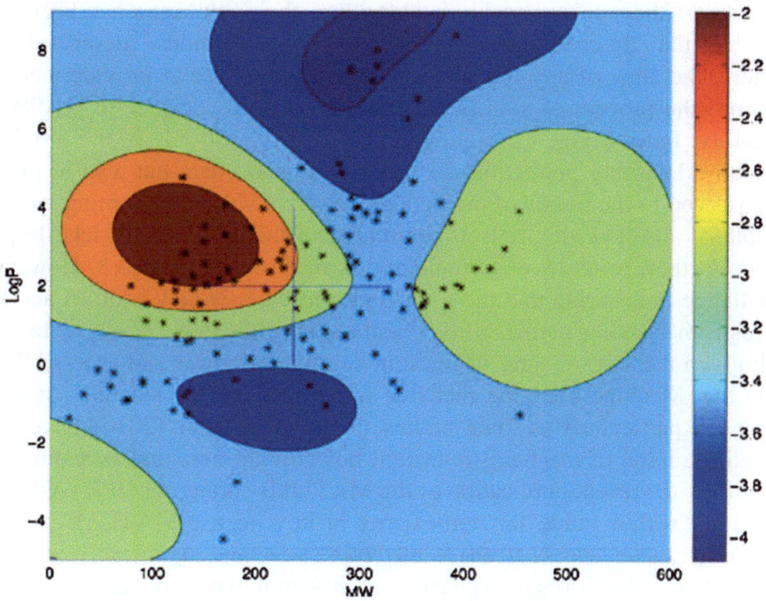

Fig. 7.6 Dependency of permeability, as log k_p, on molecular weight and lipophilicity (as log P) for a Gaussian process model of skin permeability. *Asterisks* represent training data points, and *lines* mark the median values of the descriptors shown in the plot. Modified from Sun et al. (2011)

covariance functions (specifically, *Matern3* and *Matern5* functions via MATLAB which, respectively, consider polynomials of order 1 and 2), they demonstrated little improvement in model quality in the presence of nonlinear models, and little evidence for over-fitting of data in such circumstances. However, their over-riding finding was on the nature of the data set and how it could influence the results, and any subsequent interpretation, of the resultant models. In comparative and absolute terms, the size of the data sets used for analysis significantly affect model quality and that, in order to make relevant cross-species comparisons, any model should, wherever possible, exhibit as much commonality of membership as possible.

Other researchers have also found success with nonlinear methods. For example, Neely et al. (2009) produced a statistically robust model by blending the use of quantitative structure–permeability relationships, genetic algorithms and neural networks. Their model suggested that size/shape and polarity descriptors accounted for approximately 70 % of the permeability information in their model. Brown et al. (2012) demonstrated the utility of GP-based Machine Learning methods by using them to estimate the optimum candidate for transdermal drug delivery. When compared to QSPR models, the GP methods produced better predictions of permeability and were also more able to effectively rank relative permeability for the penetrants.

The common criticisms of nonlinear models are discussed in a wider context in Chap. 9. While such methods are perceived to lack "real-world" relevance, particularly as they do not yield an algorithm, their relevance has been demonstrated by the diverse range of studies—and methods—described above. Ultimately, the aims of any models of skin absorption are to provide mechanistic insight into the process of skin permeation and to provide precise predictions of permeability. Indeed, Fig. 7.5 highlights this clearly, not only in the proximity of the Machine Learning predictions to the experimental data but also in the differences observed in the trends of predictions between Machine Learning and QSPR-type models, and how this reflects the real relevance of such models. Improving relevance further, in terms of formulation, is discussed in Chap. 8. However, the real challenge in ensuring that these methods are more widely adopted and held in higher esteem by skin permeability scientists is, ultimately, likely to be technological; given the year-on-year increase in computing power and accessibility, it is unlikely that Machine Learning methods will remain on the sidelines of this field much longer, particularly as some studies, notably Lam et al. (2010), have not only addressed the issue of mechanistic insight but brought new understandings to this field. Further, the recent publication of the MATLAB coding for GPR (Ashrafi et al. 2015), along with a guide for non-experts in how to use MATLAB-based GPR models, now makes access to these methods easier and significantly more transparent, and may lead to an expansion of research based on these methods by researchers working in a number of fields where QSAR and simple regression-based methods have tended to dominate.

References

Abraham MH, Martins F, Mitchell RC (1997) Algorithms for skin permeability using hydrogen bond descriptors: the problem of steroids. J Pharm Pharmacol 49:858–865

Agatonovic-Kustric S, Tucker IG, Schmierer D (1999) Solid state assay of ranitidine hydrochloride as a bulk drug and as active ingredient in tablets using DRIFT spectroscopy with ANNs. Pharm Res 16:1479–1484

Agatonovic-Kustrin S, Beresford R (2000) Basic concepts of artificial neural network (ANN) modelling and its application in pharmaceutical research. J Pharm Biomed Anal 22:717–727

Agatonovic-Kustrin S, Beresford R, Pauzi A, Yusof M (2001) ANN modeling of the penetration across a polydimethylsiloxane membrane from theoretically derived molecular descriptors. J Pharm Biomed Anal 26:241–254

Alany RG, Agatonovic-Kustrin S, Rades T, Tucker IG (1999) Use of artificial neural network to predict quaternary phase systems from limited experimental data. J Pharm Biomed Anal19:443–452

Ashrafi P, Moss GP, Wilkinson SC, Davey N, Sun Y (2015) The application of machine learning to the modelling of percutaneous absorption: an overview and guide. SAR QSAR Environ Res 26:181–204

Baert B, Deconinck E, van Gele M, Slodicka M, Stoppie P, Bode S, Slegers G, van der Heyden Y, Lambert J, Beetens J, de Spiegeleer B (2007) Transdermal penetration behaviour of drugs: CART-clustering, QSPR and selection of model compounds. Bioorg Med Chem 15:6943–6955

Barratt MD (1995) Quantitative structure–activity relationships for skin permeability. Toxicol Vitro 9:27–37

Bourquin J, Schmidt H, Van Hoogevest P, Leuen-Berger H (1997) Application of artificial neural networks (ANN) in the development of solid dosage forms. Pharm Dev Technol 2:111–121

Bourquin J, Schmidt H, Van Hoogevest P, Leuen-Berger H (1998) Comparison of artificial neural networks (ANN) with classical modelling technologies using different experimental designs and data from a galenical study on a solid dosage form. Eur J Pharm Sci 7:1–12

Brown SL, Rossi JE (1989) A simple method for estimating dermal absorption of chemicals in water. Chemosphere 19:1989–2001

Brown MB, Lau C-H, Lim ST, Sun Y, Davey N, Moss GP, Yoo S-H, de Muynck C (2012) An evaluation of the potential of linear and nonlinear skin permeation models for the prediction of experimentally measured percutaneous drug absorption. J Pharm Pharmacol 64:566–577

Chandrasekaran SK, Michaels AS, Campbell PS, Shaw JE (1976) Scopolamine permeation through human skin in vitro. AIChE J 22:828–832

Chen LJ, Lian GP, Han LJ (2007) Prediction of human skin permeability using artificial neural network (ANN) modelling. Acta Pharmacol Sin 28:591–600

Cronin MTD, Schultz TW (2003) Pitfalls in QSAR. J Mol Struct 622:39–51

Danick Y, Miller MA, Jaworska J, Kasting GB (2013) Design and performance of a spreadsheet-based model for estimating bioavailability of chemicals from dermal exposure. Adv Drug Deliver Rev 65:221–236

Degim T, Hadgraft J, Illbasmis S, Ozkan Y (2003) Prediction of skin penetration using artificial neural network (ANN) modelling. J Pharm Sci 92:656–664

Fan T, Takayama K, Hattori Y, Maitani Y (2004) Formulation optimisation of paclitaxel carried by PEGylated emulsions based on artificial neural network. Pharm Res 21:1692–1697

Fedors RF (1974) A method for estimating both the solubility parameters and molar volumes of liquids. Poly Eng Sci 14:147–154

Fiserova-Bergerova V, Pierce JT, Droz PO (1990) Dermal absorption potential of industrial chemicals: criteria for skin notation. Am J Ind Med 17:617–635

Flynn GL (1990) Physicochemical determinants of skin absorption. In: Gerrity TR, Henry CJ (eds) Principles of route-to-route extrapolation for risk assessment. Elsevier, New York, pp 93–127

George K (2005) A two-dimensional mathematical model of non-linear dual-sorption of percutaneous drug absorption. Biomed Eng Online 4:40

George K, Kubota K, Twizell EH (2004) A two-dimensional mathematical model of percutaneous drug absorption. Biomed Eng Online 3:18

Guy RH, Potts RO (1993) Penetration of industrial chemicals across the skin: a predictive model. Am J Ind Med 23:711–719

Keshwani DR, Jones DD, Brand RM (2005) Tagaki-Sugeno fuzzy modelling of skin permeability. Cutan Ocul Toxicol 24:149–163

Kia S, Setayeshi S, Shamsaei M, Kia M (2013) Computer-aided diagnosis (CAD) of the skin disease based an intelligent classification of sonogram using neural network. Neural Comput Appl 22:1049–1062

Kubota K, Twizell EH (1992) A nonlinear numerical model of percutaneous drug absorption. Math Biosci 108:157–178

Kubota K, Koyama E, Yasuda K (1991) A random walk method for percutaneous drug absorption pharmacokinetics: application to repeated administration of a therapeutic timolol patch. J Pharm Sci 80:752–756

Kubota K, Koyama E, Twizell EH (1993) Dual sorptionmodel for the nonlinear percutaneous permeation kinetics of timolol. J Pharm Sci 82:1205–1208

Lam LT, Sun Y, Davey N, Adams RG, Prapopoulou M, Brown MB, Moss GP (2010) The application of feature selection to the development of Gaussian process models for percutaneous absorption. J Pharm Pharmacol 62:738–749

Lien EJ, Gao H (1995) QSAR analysis of skin permeability of various drugs in man as compared to in vivo and in vitro studies in rodents. Pharm Res 4:583–587

Lim CW, Fujiwara S, Yamashita F, Hashida M (2002) Prediction of human skin permeability using a combination of molecular orbital calculations and artificial neural network. Bio Pharm Bull 25:361–366

McKone TE, Howd RA (1992) Estimating dermal uptake of nonionic organic chemicals from water and soil: I unified fugacity-based models for risk assessments. Risk Anal 12:543–557

Milik M, Kolinski A, Skolnick J (1995) Neural network system for the evaluation of side-chain packing in protein structures. Protein Eng 8:225–236

Mitragotri S, Anissimov YG, Bunge AL, Frasch HF, Guy RH, Hadgraft J, et al (2011) Mathematical models of skin permeability: an overview. Int J Pharm 418:115–129

Moss GP, Cronin MTD (2002) Quantitative structure-permeability relationships for percutaneous absorption: re-analysis of steroid data. Int J Pharm 238:105–109

Moss GP, Gullick DR, Cox PA, Alexander C, Ingram MJ, Smart JD, Pugh WJ (2006) Design, synthesis and characterisation of captopril prodrugs for enhanced percutaneous absorption. J Pharm Pharmacol 58:167–177

Moss GP, Dearden JC, Patel H, Cronin MTD (2002) Quantitative structure-permeability relationships (QSPRs) for percutaneous absorption. Toxicol In Vitro 16:299–317

Moss GP, Sun Y, Prapopoulou M, Davey N, Adams R, Pugh WJ, Brown MB (2009) The application of Gaussian processes in the prediction of percutaneous absorption. J Pharm Pharmacol 61:1147–1153

Moss GP, Sun Y, Wilkinson SC, Davey N, Adams R, Martin GP, Prapopoulou M, Brown MB (2011) The application and limitations of mathematical models across mammalian skin and poldimethylsiloxane membranes. J Pharm Pharmacol 63:1411–1427

Neely B, Madihally S, Robinson RJ, Gasem K (2009) Nonlinear quantitative structure–property relationship modeling of skin permeation coefficient. J Pharm Sci 98:4069–4084

Pannier AK, Brand RM, Jones DD (2003) Fuzzy modelling of skin permeability coefficients. Pharm Res 20:143–148

Patel H, ten Berge W, Cronin MTD (2002) Quantitative structure-activity relationships (QSARs) for prediction of skin permeation of exogenous chemicals. Chemosphere 48:603–613

Potts RO, Guy RH (1992) Predicting skin permeability. Pharm Res 9:663–669

Pugh WJ, Roberts MS, Hadgraft J (1996) Epidermal permeability—penetrant structure relationships 3: the effect of hydrogen bonding interactions and molecular size on diffusion across the stratum corneum. Int J Pharm 138:149–165

Roberts MS, Pugh WJ, Hadgraft J, Watkinson AC (1995) Epidermal permeability-penetrant structure relationships 1: an analysis of methods of predicting penetration of monofunctional solutes from aqueous solutions. Int J Pharm 126:219–233

Roberts MS, Pugh WJ, Hadgraft J (1996) Epidermal permeability: penetrant structure relationships 2: the effect of H-bonding groups in penetrants on their diffusion through the stratum corneum. Int J Pharm 132:23–32

Roberts MS, Anissimov YG, Gonsalvez RA (1999) Mathematical models in percutaneous absorption. In: Bronaugh RL, Maibach HI (eds) Percutaneous absorption, 3rd edn. Marcel Dekker, Inc. CRC Press, New York, pp 3–56

Roberts MS, Cross SE, Pellet MA (2002) Skin transport. In: Walters KA (ed.) Dermatological and transdermal formulations. Marcel Dekker, New York, pp 89–195

Robinson PJ (1993) A composite model for predicting dermal penetration in vivo. In: Human and environmental safety division. The Proctor & Gamble Company, Cincinati. Personal communication, cited in Wilschut A, ten Berge WF, Robinson PJ, McKone TE (1995) Estimating skin permeation: the validation of five mathematical skin permeation models, vol 30. Chemosphere, Los Angeles, pp 1275–1296

So SS, Karplus M (1996) Evolutionary optimization in quantitative structure-activity relationship: an application of genetic neural networks. J Med Chem 39:1521–1530

So SS, Karplus M (1997a) Three-dimensional quantitative structure-activity relationships from molecular similarity matrices and genetic neural networks 1: methods and validations. J Med Chem 40:4347–4359

So SS, Karplus M (1997b) Three-dimensional quantitative structure-activity relationships from molecular similarity matrices and genetic neural networks 2: applications. J Med Chem 40:4360–4371

Sun Z, Rao X, Peng L, Xu D (1997) Prediction of protein supersecondary structures based on the artificial neural network method. Protein Eng 10:763–769

Sun Y, Moss GP, Prapodopolou M, Davey N, Adams R, Brown MB (2008) Predictions of skin penetration using machine learning methods. In: Giannotti F, Gunopulos D, Turini F, Zaniolo C, Ramakrishnan N, Wu XD (eds) Proceedings of 8th IEEE international conference on data mining, (ICDM 2008). IEEE, Pisa, ISBN 978-0-7695-3502-9

Sun Y, Moss GP, Davey N, Adams R, Brown MB (2011) The application of stochastic machine learning methods in the prediction of skin penetration. Appl Soft Comput 11:2367–2375

Takahara J, Takayama K, Nagai T (1998) Multi-objective simultaneous optimization technique based on an artificial neural network in sustained release formulations. J. Cont. Relat 49:11–20

Willett P (1995) Genetic algorithms in molecular recognition and design. Trends Biotechnol 13:516–521

Williams AC (2003) Transdermal and topical drug delivery. The Pharmaceutical Press, London

Wilschut A, ten Berge WF, Robinson PJ, McKone TE (1995) Estimating skin permeation. In: The validation of five mathematical skin permeation models, vol 30. Chemosphere, Los Angeles, pp 1275–1296

Wu CH (1997) Artificial neural networks for molecular sequence analysis. Comput Chem 21:237–256

Chapter 8
Finite-Dose Models of Transient Exposures and Volatile Formulation Components

Introduction

The majority of the work described so far, irrespective of mathematical or computational methods employed, has been modelled using data from infinite-dose experiments. This technique is described in Chap. 2 and was used in the earliest types of studies in this field (Flynn 1990; El Tayar et al. 1991; Potts and Guy 1992) as it has significant relevance to the manner in which potential permeants may become available for absorption into and across the skin, as discussed in Chap. 3. Flynn discussed skin permeability in the context of pollutants and their risk to human health, commenting that the majority of such pollutants will be bioavailable via exposure to the skin of aqueous solutions and that such systems therefore define the upper limit on absorption of a chemical across the skin. Further, the solvent—which can exert changes to the skin barrier properties—is central to considerations of permeability in such models, particularly during situations where the skin exposure occurs over a long duration. This, as with all in vitro models of in vivo processes, is developed in this context and with its limitations clearly defined.

There are, however, many situations where an "infinite dose" is not presented to the skin. These may be pharmaceutical, where dosage forms such as transdermal patches and topically applied creams and ointments will essentially present a finite amount of the prospective penetrant to the skin over a given time. They may, for example, also deal with the absorption of industrial chemicals in any number of scenarios, from the handling of bulk industrial chemicals to the spraying of pesticides onto crops, or to situations where multiple, minor exposures are experienced. In such circumstances, the application of a dose to the skin surface is characterised by a finite application, or a number of finite applications, often with a solvent that is transiently in contact with the skin due either to environmental factors—including its lack of occlusion, or its volatility. Steady-state permeation is unlikely to be achieved in such circumstances.

G.P. Moss et al., *Predictive Methods in Percutaneous Absorption*,
DOI 10.1007/978-3-662-47371-9_8

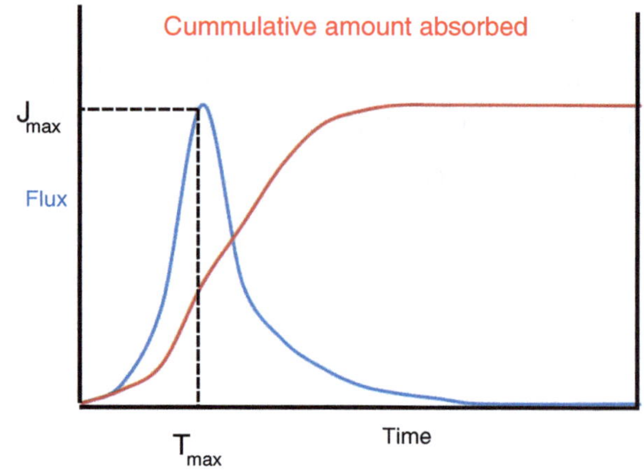

Fig. 8.1 Schematic representation of flux and cumulative absorption profiles for a finite-dose application to human skin (© Williams (2003), used with permission)

This is shown in Fig. 8.1, which plots schematically flux and cumulative absorption profiles for a finite-dose experiment. Initially, the flux (J_{max}), measured throughout the experiment and often reported as the *instantaneous flux*, will increase and reach a maximum (T_{max}) at an early stage of the process, and it will decrease thereafter—this decrease in flux may be associated with the depletion of the donor phase, or the amount of material for absorption, remaining on the skin surface, or formulation effects, which might include solvent evaporation which may result in the potential penetrant remaining on the skin surface as a solid, unable to permeate into the skin in the absence of its solvent. Thus, the cumulative amount absorbed will increase and then reach a plateau unless further doses are applied to the skin surface, which may be the case for chemicals in pharmaceutical or various industrial applications and even in cosmetic applications where repeat dosing is common on a regular (e.g. daily) basis. The shape of the flux profile can yield interesting information and, for example, may, in the case of peak broadening, be indicative of the extent of binding of the penetrant in the membrane and therefore be suggestive of retention of a penetrant in the skin, possibly reflecting the "reservoir effect" often associated with the *stratum corneum*.

Designing in vitro experiments that accurately reflect this process may be difficult. Application of materials to the skin surface in volatile solvents may result in damage to the skin barrier, usually by delipidisation. Such damage is normally minimised by the use of volatile solvents which are removed rapidly from the skin surface. This will also result in a thin film of solid solute being deposited on the skin surface. The use of a volatile liquid solvent to achieve this results in greater accuracy in dosing as it removes issues associated with gravimetric dosing, being easier to administer to small diffusion cells, and also allows the dose to be more evenly applied to the skin surface (Williams 2003). It follows from such a discussion that the development of quantitative models describing such processes will be equally challenging.

Infinite-dose profiles usually report the steady-state flux of the absorption process, or its derivative, k_p, the permeability coefficient. In the case of finite-dose experiments, the maximum flux (J_{max}) and the time to reach maximum flux (T_{max}) are often reported. Crank (1975) showed that J_{max} may be defined as:

$$J_{max} = \frac{1.85DC_0\delta}{h^2} \tag{8.1}$$

where

- D is the apparent diffusion coefficient;
- C_0 is the concentration of the permeant in the first layer of the *stratum corneum*, which is maximal when solid deposited drug is in contact with the skin surface;
- h is the thickness of the *stratum corneum*;
- δ is the thickness of the finite-dose layer on the skin surface.

T_{max} may be represented by the expression:

$$T_{max} = \frac{h^2 - \delta^2}{6D} \tag{8.2}$$

where the terms used in Eq. 8.2 are as defined for Eq. 8.1.

As a finite dose, δ, is significantly smaller than the thickness of the *stratum corneum*, Eq. 8.2 is often simplified to:

$$T_{max} = \frac{h^2}{6D} \tag{8.3}$$

These equations allow the estimation of the apparent diffusion coefficient, although any such value should be calculated in the context of the limitations of the finite-dose model described above.

Franz et al. (1993) illustrated the differences in finite- and infinite-dose systems by comparing the permeation of *N*-nitrosodiethanolamine (NDELA) through human skin in vitro. Their results, for a range of cosmetic vehicles, are summarised in Table 8.1. They show that for an infinite dose of NDELA (500 μg/cm^2 at 1 μg/μl),

Table 8.1 Percutaneous penetration of *N*-nitrosodiethanolamine (NDELA) through human skin in vitro for a range of cosmetic formulations in both finite- and infinite-dose systems

	Infinite dose (500 μl/cm^2)	Finite dose		
		10 μl/cm^2	10 μl/cm^2	10 μl/cm^2
Total dose (μg/cm^2)	500	0.6	2.0	6.0
Concentration (μg/μl)	1.0	0.06	0.20	0.6
IPM	24.9 ± 5.2	41.8 ± 10.9	65.2 ± 9.5	61.9 ± 13.9
Shampoo	35.4 ± 7.6	42.7 ± 15.1	42.9 ± 13.7	46.4 ± 12.4
Sunscreen	3.9 ± 1.6	37.3 ± 12.3	35.3 ± 8.3	41.4 ± 14.5

Modified from Franz et al. (1993)

the total dose absorbed was formulation dependant. There was also evidence that the rate of absorption increased with time which the authors claimed suggested skin barrier damage. Three finite doses were applied (0.6, 2.0 and 6.0 μg/cm^2, applied as 10 μl of 0.06, 0.2 and 0.6 μg/μl, respectively) to the skin, and the percentage absorption was found to be largely concentration independent as the total dose absorbed (as μg/cm^2) was directly proportional to the concentration of NDELA in the vehicle. As with infinite-dose studies, the total absorbed at the end of the 48-h experiment was formulation dependent. Table 8.1 shows a comparison of the infinite- and finite-dose experiments, with absorption presented as a percentage of the applied dose. For all formulations, it is apparent that the finite-dose experiments result in greater percentage permeation.

As such experiments provide experimental data for modelling studies, there has emerged a body of research that has focused on the use of finite-dose models in characterising skin permeation in a range of systems (for pharmaceutical, bulk industrial, cosmetic and risk assessment/toxicity purposes). The rest of this chapter will discuss such models.

Modelling Finite-Dose Experiments

Finite-dose systems are usually associated with transient or unoccluded applications of small amounts to the skin surface. This may occur in a range of contexts, from pesticide application to crops, to industrial exposures associated with volatile solvents or the topical application of unoccluded medicines such as creams or ointments. A number of studies have examined dermal absorption under such circumstances and have provided a platform from which models have been subsequently developed. For example, a number of studies on the dermal absorption of aviation fuel have indicated that absorption depends on the type of fuel and its specific components, with some materials resulting in different peak absorptions; for example, Riviere et al. (1999) determined that naphthalene had a peak flux 1 h following a transient exposure whereas dodecane and hexadecane had prolonged, but lower, fluxes. A subsequent study found that there are no significant differences in the ability of four different fuels to influence absorption (Baynes et al. 2000).

Other topically applied volatile chemicals have also been examined. Riviere et al. (2000) examined the absorption of chloropentafluorobenzene and dichlorobenzene, suggesting that dosing methodology is a major factor in the degree of subsequent dermal absorption and that methodological factors could influence measurement of such uptake. As "real-world" formulations applied to the skin are seldom simple, several studies have explored the effect of solvent mixtures on skin absorption. Baynes et al. (2001) investigated mixture effects on the dermal absorption of aviation fuel components. They found, after a series of experiments which examined systematic variations in the composition of performance additives, that such changes significantly altered the dermal disposition of aromatic and aliphatic fuel components within the skin. They also found that the products of two-factor interactions could not be predicted from two, single-factor exposures. Riviere et al. (2001) examined the influence of mixture effects on the dermal absorption of pentachlorophenol and concluded that dermal absorption estimates from single chemical exposures do not reflect the dermal absorption measured following application of chemical mixtures. This theme is discussed in the following section in more detail.

Frasch and Barbero (2008) considered several examples of dermal absorption following transient exposure, using infinite-dose experiments (permeability and lag time) to predict the accumulation of chemicals in the skin following transient exposure and by applying Laplace domain solutions to their theoretical framework. They considered the skin as a homogenous membrane which is free of chemicals at the beginning of the experiment, after which time a chemical of known concentration (C_1) is applied for a finite time (T_1), after which it is completely removed. In using the one-dimensional diffusion equation:

$$\frac{\partial C}{\partial t} = D\frac{\partial^2 C}{\partial x^2} \tag{8.4}$$

where
D is diffusivity;
t is time;
x is position.

Frasch and Barbero then defined the flux as:

$$f(x,t) = -D\frac{\partial C}{\partial x} \tag{8.5}$$

Equation 8.5 is then integrated with respect to t, which gives the total mass accumulated per unit area after time t:

$$m(t) = \int_0^t f(h,t)\mathrm{d}t \tag{8.6}$$

They then considered a number of exposure conditions. These included the situation of zero concentration at the upper surface (the outer surface of the skin). In this scenario, the concentration at the surface of the skin is maintained at zero, which reflects an infinite, well-stirred reservoir on the skin surface for a highly volatile compound. In this case, the chemical diffuses upwards through the skin and immediately evaporates when it reaches the skin surface, being carried into the surrounding environment. In this case, the total mass accumulation is defined as:

$$m_\infty = k_p \sum_{i=1}^{n} A_i C_i T_i \tag{8.7}$$

where

m_∞	is the total mass accumulation as time approaches infinity;
$k_p = K_{mv}$. D/h	the permeability coefficient, where K_{mv} is the membrane–vehicle partition coefficient;
A_i	is the area of skin exposed to the chemical;
C_i and T_i	relate to multiple intermittent exposures of the skin of concentration C_i for duration T_i.

They also considered the scenario where zero flux was observed at the upper surface of the skin. Following the exposure period, the upper skin surface presents an impermeable barrier to diffusion and represents, in this model, the potential permeation of a non-volatile compound where absorption into the skin is energetically preferred to evaporation into the surrounding environment. In this case, the total mass accumulation as time approaches infinity is represented by the sum of the amount in the membrane at time T_i:

$$m_m = \frac{AhC_1}{2}\left[1 - \frac{8}{\pi^2}\sum_{n=0}^{\infty}\frac{1}{(2n+1)^2}\exp(-D(2n+1)^2\pi^2\frac{T_1}{h^2}\right] \tag{8.8}$$

where

T_1 is the amount that has passed through the membrane at time t;

h is the thickness of the membrane.

The sum of the mass, rewritten in terms of permeability coefficient and lag time, is:

$$m_\infty = k_p A C_1\left[T_1 + 2\tau - \frac{12\tau}{\pi^2}\sum_{n=1}^{\infty}\frac{1}{n^2}\exp\left(-\frac{n^2\pi^2 T_1}{6\tau}\right)\right] \tag{8.9}$$

where $\tau = h_2/(6D)$, the membrane lag time.

For long exposure times, the exponential component of Eq. 8.9 becomes negligible, allowing the approximation shown in Eq. 8.10 to be used as:

$$m_\infty \approx k_p A C_1 (T_1 + 2\tau) \tag{8.10}$$

And, for very long exposure times, such as when T_1 is significantly greater than 2τ, Frasch and Barbero suggested that Eq. 8.10 could be simplified further to:

$$m_{\infty} \approx k_p A C_1 T_1 \tag{8.11}$$

Comparison with Eq. 8.7 suggests that the total mass accumulation is the same in both scenarios and that a similar situation may occur when the lag time is very small.

They then applied their theoretical considerations to the in vitro permeation of diethylphthalate applied to hairless guinea pigskin and to silicone membranes as transient exposures. In general, they found excellent correlations between their theoretical framework and the experimental results, with r^2 values ranging from 0.97 to 0.999. Even though the discrepancies between theoretical estimates and practical observations are very minor, the authors considered the use of silicone membranes to explore issues with skin heterogeneity, suggesting that transient dose experiments supported to a degree the homogenous membrane theory proposed in their theoretical framework along with other factors, such as the finite time required to wash membranes which, while kept as brief as possible, may still have resulted in some diffusion during the process, or the potential for their model penetrant (diethylphthalate) to bind to skin components.

A two-stage kinetic analysis was adopted by Saiyasombati and Kasting (2003) to characterise fragrance evaporation and absorption following topical application to the skin. Their model considered low doses—typically associated with the topical application of fragrances from a range of vehicles—and used a previously published data set containing evaporation rates of topically applied volatile chemicals. Evaporation rates from formulations containing a musk fixative (a commonly used ingredient to retard evaporation of substances from the skin surface and increase the life of a fragrance on the skin) followed single exponential decays that could be described by a single-compartment model. In contrast, the same system without the musk fixative resulted in two-compartment biexponential decay. None of the models presented accounted for the ingredient interactions that lead to the observed differences, which led the authors to suggest the adoption of a diffusion/evaporation model in subsequent studies. Nevertheless, their compartmental kinetic models yielded reasonable correlations with the total evaporated fraction of each component in the formulations evaluated and implied a relationship, in the absence of skin binding or chemical reactivity, between molecular properties of an ingredient and the fraction of the dose absorbed. They concluded that the two-compartment model, where evaporation from the vehicle and skin was both considered, provided the most useful model in quantifying the duration of the fragrances on the skin.

Miller et al. (2005) investigated the effect of dose and airflow dependence on the disposition of benzyl alcohol on the skin. They showed that the dose absorbed after 24 h increased with the amount applied to the skin for nine different doses. Less than 4 % of the radiolabelled active was retained in the tissue after this time and the rest was assumed to have evaporated. Analysis of their data showed that the increase in benzyl alcohol absorption was consistent with an increase in diffusivity in the *stratum corneum*, as the concentration increased. Their variable diffusivity model (Eq. 8.12) was able to estimate the observations to within 4.2 % of dose:

$$D = D_0 + \frac{D_{\text{sat}} - D_0}{1 + \exp\left[m\left(1 - \frac{C}{C_{\text{trans}}}\right)\right]} \tag{8.12}$$

where

D is the diffusivity of the permeant in the membrane,

D_0 and D_{sat} reflect the concentration-dependent diffusion coefficient which varied from approximately D_0 at low concentrations to D_{sat} at the saturation concentration, C_{sat},

$C = C(x)$ the local concentration of permeant in the membrane.

They concluded that the experimentally measured dose- and airflow-dependant absorption and evaporation of benzyl alcohol in vitro fit well to their proposed model and were consistent with a finite-dose diffusion model describing evaporative loss from the skin surface and concentration-dependant diffusion from the skin surface. In addition, while Miller et al. discussed the relevance and validity of their predictive model, they also highlighted its application to a limited number of chemicals.

Kasting and Miller (2006) proposed models for the disposition of volatile compounds to the skin surface. They examined two cases: the first was where the dose was less than that required to saturate the skin surface, and the second was where absorption and evaporation approach steady state as the dose is increased. In the first case, where the applied dose is less than or equal to the saturation dose, they described *stratum corneum* transport in the context of one-dimensional diffusion. In the second case, where the applied dose is greater than the saturation dose, and where it is assumed that evaporation from the skin surface occurs at a constant rate, they describe the deposition of a dose (M_{sat}) into the upper layers of the *stratum corneum*. The dose remaining on the surface ($M_{\text{surf}} = M_0 - M_{\text{sat}}$) forms a residual surface reservoir, which replenishes the top layers of the *stratum corneum* as absorption into lower layers occurs; this residual surface layer is itself prone to evaporation until depleted. Permeant transport in the membrane is governed by the one-dimensional diffusion equation (Eq. 8.4) with initial constraints which consider permeant depletion. Kasting and Miller comment that their models are applicable to simple solvent-deposited liquids and solids but are not able to describe complex mixtures and provide an approximate estimate of absorption following transient skin contact. They also described the complexity of the process of finite-dose absorption, which typically include a wide range of experimental conditions, complex and dynamic solution thermodynamics during the drying phase following deposition, interaction of formulation components with the skin (potentially altering its structure), complex mathematics describing the non-steady state and, finally, a shortage of experimental data on which to test and develop their theoretical framework. Nevertheless, their diffusion models offer improved predictions compared to compartmental models for skin absorption, as the latter are hindered by only being able to consider small doses and whose predictive accuracy might be limited by the use of approximations associated with the use of particular physical properties which make it difficult to estimate absorption compared to the simple one-compartment model.

Diffusion-based mathematical models of dermal permeation were considered by Kruse et al. (2007). They measured permeability for a series of finite- and infinite-dose experiments for a data set of varying lipophilicities. They defined three dermal permeation processes which could be differentiated and defined: exposure by contact with the outer skin area, diffusion-controlled partition through the various skin layers and uptake in the systemic circulation in vivo (or passage into the receptor compartment of an in vitro diffusion experiment). They thus employed a model based on this which had four compartments—the donor phase, the *stratum corneum*, the viable epidermis and the receptor phase (characterised as appropriate by diffusion coefficients and the thickness of the skin layers) with the distribution between layers determined by partition coefficients. They applied the Kruse model to their data to describe mass flows between compartments and the finite- and infinite-dose models reported by Anissimov and Roberts (1999, 2001). The models were further described by the use of occupationally relevant descriptors, such as, for example, the wipe variable (Q_{wiped}) which relates to the amount of an applied dose removed from the skin at the termination of exposure(s), and also by the physicochemical parameters used in other models of skin absorption (Potts and Guy 1992; Cleek and Bunge 1993; Bunge and Cleek 1995). Application of solids was understood to initially involve their dissolution into a vehicle present on the skin surface, such as sweat, and their aqueous solubility was assumed to determine the maximum concentration in the donor compartment in contact with the skin, and therefore the maximum steady-state rate. The models developed demonstrated good fits and, in a variety of exposure environments, good agreement. In attempting to predict the amount absorbed following a finite-dose application, the time course of cumulative mass absorbed for the duration of the experiment was estimated. Overall, it was found that the models developed were able to predict reasonably well permeation up to approximately 6 h. Thereafter, the models did not perform as well, which was probably due to the changing nature of the applied dose due to, for example, evaporation from the skin surface or a change in state of components of the applied formulation. For a finite dose, where finite-dose data are available, they proposed the following algorithm:

- Assume infinite and finite dose data to begin with.
- Fit the model to infinite dose data to obtain values for $K_{sc/d}$ [the *stratum corneum*/donor partition coefficient, which may be written as $K_{sc/w}$ when the donor is aqueous] and D_{sc} [the diffusion coefficient in the *stratum corneum*]. It must be assumed that at least two solutions will be obtained for the parameter values.
- Predict finite dose time course which matches finite dose data available for each set of parameter values obtained from fitting.
- Compare predicted time course for each parameter set with the experimental time course. Choose a single parameter set based on this comparison.
- Make further predictions with model and the chosen parameter set. Limit all predictions to less than six hours after the start of exposure.

(from Kruse et al. 2007)

In essence, Kruse et al. have developed a simple model that extends the use of data from infinite-dose data to the prediction of finite-dose exposures. They used this model to provide an accurate estimate of finite-dose, or transient, exposure for periods up to 6 h. Thereafter, the model is less capable of estimating exposure, mostly due to physical changes to the applied dose. They also observed that, for one lipophilic chemical (triclosan), the predicted permeability was significantly larger than the measured permeability, perhaps reflecting the underlying context of linear QSAR-type models of skin permeability.

Other researchers have also explored the development of finite dose. Buist et al. (2010) developed finite-dose models based on infinite-dose models (Potts and Guy 1992; Cleek and Bunge (1993); Patel et al. 2002) and the *stratum corneum*/water partition coefficient which were relevant to non-volatile substances dissolved in water at non-irritating concentrations, specifically by using their permeability coefficients (k_p) and lag times as the basis of their model. The model was tested by comparing its output to the results obtained from a series of in vitro experiments. In general, while they found that the model in most cases over-predicted, measured absorption predicted values were either not statistically different from them or over-predicted them. In the latter case, over-predictions were still below European default absorption values (EU 2004).

With an echo of the applicability of complex mathematical models to wider fields, including those who may not necessarily be subject experts, Dancik et al. (2013) developed an excellent and widely applicable one-dimensional spreadsheet-based model for estimating bioavailability of chemicals from dermal exposure. The model details the structure of the permeability strata and treats the viable epidermis as an unperfused dermis and is formulated from first-principles-based transport equations and empirical partition and diffusion coefficients. They thus developed, in an easily used format, a complex model that found excellent agreement with experimental findings. While the model addresses key issues in applicability and use, they also comment that its underlying complexity makes it difficult to infer detailed mechanistic information from the model.

Models of Formulation in Finite-Dose Experiments

It should be noted that, following the original study by Flynn and the adoption of his data set, or variations thereon, the majority of work in this field has focused on the development of models of permeability from saturated aqueous solutions. Riviere and Brooks (2005) developed an approach, using flow-through diffusion cells, whereby solvent mixtures for twelve penetrants were used to develop QSARs which addressed the effect of solvents other than water on skin permeability—in effect, as this examined bulk properties of the solvents, this was addressing non-specific issues such as variations in solvent polarity and differences between the affinity of particular solvents and solutes. Thus, for a total of 344 combinations, they found statistically significant QSARs which related permeability to refractive index, polarisability and the Henry's Law Constant. The authors thus presented a method for quantifying the effect of solvents on skin permeability, which is expanded upon in subsequent studies discussed below. For example, Riviere and Brooks (2007) used hybrid QSARs to examine the effect the mixture factor of solvents had on skin permeation of model penetrants (12 compounds and 24 mixtures, for a total of 288 distinct systems). They found that this approach was able to yield improved QSARs of complex chemical mixtures.

Ghafourian et al. (2010a) developed QSARs for finite-dose systems where formulation effects were considered; commonly, such effects, particularly those associated with solvents or chemical penetration enhancers (e.g. ethanol, propylene glycol, urea and dimethylsulphoxide) are widely known to reduce the *stratum corneum* barrier function and increase permeability across the skin. Ghafourian et al. (2010a) investigated the permeability of four chemicals from a range of 24 solvent blends, giving a total of 96 permeability coefficients, in a finite-dose in vitro model using a porcine skin membrane. These were added to a previous data set of 288 permeability (k_p) values (taken from Riviere and Brooks 2005), resulting in a data set comprising 384 unique measurements of k_p for 16 different chemicals in a range of solvents and solvent mixtures. A range of molecular descriptors was determined using several software packages, including a predicted value for log *P*. For the mixtures of solvents used, averages of physicochemical properties were calculated using the fractions of each component in a mixture. Stepwise regression analysis was performed with log k_p as the dependant variable and the chosen molecular descriptors (which were summarised in the paper but not explicitly listed) for various sets of descriptors, thus generating a number of QSARs describing the permeability process for each subset. They limited the number of descriptors used in each analysis to four, in order to reduce the risk of chance correlations. Their analysis yielded four QSARs which described permeability:

$$\log k_p = -0.956 - 0.00322\Delta\text{mp} - 0.00320W(P) - 0.0121\text{BP}(V) - 0.114\text{Lipole}(P)$$
$$[n = 384 \quad s = 0.478 \quad r^2 = 0.701] \tag{8.13}$$

$$\log k_p = -310 - 0.000315W(P) - 0.00771\delta(V).E_{\text{HOMO}}(P) - 0.0102\text{BP}(V) - 0.0750\text{Lipole}(P)$$
$$[n = 384 \quad s = 0.494 \quad r^2 = 0.681] \tag{8.14}$$

$$\log k_p = -2.48 - 0.0474N_{\text{atoms}}(P) - 0.00798\delta(V).E_{\text{HOMO}}(P) - 0.0102\text{BP}(V) - 0.0723\text{Lipole}(P)$$
$$[n = 384 \quad s = 0.516 \quad r^2 = 0.653] \tag{8.15}$$

$$\log k_p = -4.29 - 0.0474N_{\text{atoms}}(P) - 0.00904\text{BP} - \text{MP}(V) - 0.345E_{\text{HOMO}}(P) - 0.0790\text{Lipole}(P)$$
$$[n = 384 \quad s = 0.522 \quad r^2 = 0.644] \tag{8.16}$$

where

mp	is the difference between the melting point of the penetrant and the melting point of the solvent;
W	is the Weiner topological index, which is the sum of distances between all pairs of vertices in the molecular graph of an alkane (defined by Diudea and Gutman 1998);
δ	is the Hildebrand solubility parameter;
E_{HOMO}	is the energy of the highest occupied molecular orbital;
BP	is the boiling point;
N_{atoms}	is the total number of atoms in the molecules;
BP–MP	is the difference between the boiling point and the melting point of a compound;
Lipole	is the total lipole moment of a chemical;
(*P*) and (*V*)	represent penetrant and vehicle phenomena/descriptors, respectively.

Each equation contains 2–3 penetrant descriptors and 1–2 vehicle descriptors and, in the case of Eqs. 8.13–8.16, one combined vehicle–penetrant descriptor.

The N_{atoms} and W descriptors characterise molecular size (Diudea and Gutman 1998). In all cases, the above equations highlight that increasing size decreases permeability. Further, *lipole*, a descriptor characterising the distribution of lipophilicity on a molecule calculated from local values of log P within a molecule (Pedretti et al. 2002), also exerts a negative influence on permeability. This is often associated with directional lipophilicity within a molecule, such as that associated with surfactants which normally have clear polar and non-polar ends. Therefore, Ghafourian's QSARs suggest that less polar, or more specifically *lipolar*, penetrants will have higher permeability. Nucleophilicity, represented by E_{HOMO}, a term describing the highest occupied molecular orbital, was also found to be significant, suggesting that highly nucleophilic compounds (usually those containing aromatic rings) have the lowest permeability. In Eqs. 8.14 and 8.15, the product of vehicle and penetrant terms, i.e. the product of E_{HOMO} and vehicle solubility, δ, suggests that highly nucleophilic penetrants will be less permeable if delivered from vehicles that are highly associated (i.e. vehicles that feature high intermolecular interactions, such as those in which hydrogen bonding is common). The vehicle descriptor most commonly found in their QSARs was boiling point, which again exerted a negative effect on permeability. Along with the solubility parameter, which was found in some models, these terms are interpreted by the authors as representing the intermolecular interaction energy of the vehicle, which is an effect normally associated with the polarity of the vehicle constituents. Their models suggest that reducing vehicle polarity increases the rate of skin permeation. Similarly, the significance of the solubility parameter and E_{HOMO} parameters is discussed in the context of polarity. The difference in the melting points of the vehicle and penetrant is highly significant in Eq. 8.13 and suggests that the melting points of both should be as close as possible to improve skin permeation, which the authors highlighted sits with previous observations on the permeation and melting point of isomers (Cilurzo et al. 2010).

Thus, the BP–MP(V) descriptor which finds significance in Eq. 8.16 quantifies this phenomenon, and Ghafourian et al. (2010a) discuss this in the context of similarities and differences in the molecular symmetry of the solvents and the solutes (permeants). In particular, they highlight that propylene glycol has the highest difference between it and its solutes, reducing skin permeability from such vehicles. Of the four equations, Gahfourian describes it is Eq. 8.13 which is the statistically most robust, having the highest correlation and the lowest error, suggesting that this equation is the most significant of the four they present.

Thus, Ghafourian et al. (2010a) provide a systematic and detailed investigation of the effects of solvents on the skin permeability and quantify this using a QSAR approach. They are able to relate their outputs to specific molecular phenomena which have a real-world application in advising on formulation design to optimise drug delivery. Similarly, these authors also developed similar QSAR methods to model solvent effects on the permeation of a similar data set to that described above (Ghafourian et al. 2010b). The study design was similar to their previous study (Ghafourian et al. 2010a), using similar diffusion apparatus and a porcine membrane, but also considered established models of skin permeability. Stepwise regression resulted in the following QSAR:

$$\begin{aligned} &\log k_p = -0.909 - 0.610 \log P + 2.62_p - 0.00918(\mathrm{SolBP} - \mathrm{SolMP}) \\ &\left[n = 288 \quad S = 0.438 \quad r^2 = 0.729 \quad F = 255.2 \quad P = 0.000\right] \end{aligned} \tag{8.17}$$

where

log P	is the octanol–water partition coefficient;
χ_p	is the 9th order path molecular connectivity index;
SolBP–SolMP	is the difference between the boiling point and the melting point of the solvent system.

This expression features two penetrant descriptors and one solvent mixture descriptor. Log P was the most significant descriptor, but it is interesting to observe that in Eq. 8.8, it has a negative effect on skin permeation unlike the vast majority of other QSAR models—this may be an artefact of the different experimental protocols used in different studies, such as the use in these studies of finite-dose methods. The authors attribute the effect observed in their study to the lipophilic nature of their model compounds which, while still within the same "molecular space" as common QSAR models, may be somewhat skewed towards a higher mean log P value. Such observations also sit with the considerations addressed in Chaps. 7 and 9 in relation to the use of nonlinear models, where increases in log P beyond 2.5–3.0 result in a decrease in permeability, and more broadly in the consideration of nonlinear (e.g. parabolic or bilinear) nature of biological responses observed in other fields (Hansch and Clayton 1973; Kubinyi 1977). The χ_p descriptor relates to the presence of chains of nine atoms in a molecule and in this study is related by the authors to an ideal molecular weight for skin permeation of approximately 350 Da. Finally, the difference in melting and boiling points is again significant in the context of molecular symmetry described above and detailed elsewhere (Slovokhotov et al. 2007). As in the previous studies, similar findings for the effect of propylene glycol in reducing skin permeability were also found.

Comparison was also made to the output of this study to previous QSAR studies (Flynn 1990; Potts and Guy 1992; Wilschut et al. 1995) by a range of methods, including principal component analysis (PCA) involving all 128 descriptors computed for each of the penetrants used in this study, with the emphasis on data similarity across the different studies. Finally, focus was given to those descriptors mostly associated from previous studies with skin absorption and determined, in this study, by stepwise regression. Thus, the following QSAR was derived:

$$\begin{aligned} &\log k_p = -2.91 + 0.62 \log P + 5.21 \cdot (10)\chi_p^v - 1.64 \cdot (6)\chi_p^v \\ &\left[n = 139 \quad S = 0.548 \quad r^2 = 0.757 \quad F = 140 \quad P = 0.000\right] \end{aligned} \tag{8.18}$$

where

$\log P$ is the octanol–water partition coefficient;

$(10)\chi_p^v$ is the 10th-order valence-corrected path molecular connectivity index;

$(6)\chi_p^v$ is the 6th-order valence-corrected path molecular connectivity index.

Both $(10)\chi_p^v$ and $(6)\chi_p^v$ are topological features that describe the occurrence in a molecule of certain chemical fragments; in this case, they represent 6- and 10-atom unbranched chains present in molecules, highlighting their significance to the model derived; it is interesting to note that the 6th-order descriptor has a negative influence, whereas the 10th-order descriptor has a positive influence on permeability.

Riviere and Brooks (2011) explored the influence of skin biology on the utility of QSPR models. Again, they focused on mixtures applied to porcine skin in vitro and developed QSPRs from the resultant permeability data. They used 16 chemicals and a range of formulations, which resulted in a total of 384 treatments in a flow-through diffusion cell, and also examined 20 chemicals in 119 formulation treatments in an isolated perfused skin model. They fit their data to a modification of a previously published model (by Abraham and Martin 2004, to which was added a sixth descriptor which accounts for mixture effects, as described previously by Riviere and various co-authors). Correlation of experimental and predicted data was good (r^2 values ranging from 0.69 to 0.73). A range of mixture factors were required for each model system. Their approach meant that the number of descriptors from the previously published method (five) could be reduced by analysing the data by including a mixture term and that the terms removed from the original models depended on the experimental protocol used. Thus, they developed a QSPR which describes the effect of solvent mixtures and which was dependent upon the methodology employed to produce the data. Further, in discussing the modelling approach and the use of methods that address the presence of mixtures, they describe the known parabolic relationship between skin permeability and log P, which increases to a plateau at log *P* 2–3 and decreases thereafter, and that this plateau is formulation dependent; for example, the plateau moves to a lower log P value when lauric acid is added to an aqueous formulation (Lee et al. 1994). This, Riviere and Brooks claim, agrees with their mixture-factor concept and how it influences permeability and, in this context, how it improves the accuracy and relevance of a model of percutaneous absorption.

In both of their experimental arrangements, Riviere and Brooks used an albumin-containing perfusate which is known to facilitate the absorption of lipophilic compounds (Bronaugh and Stewart 1984; Cross et al. 2003). They suggest that this modification to their receptor compartment alters the dermal penetration profile in a manner that removes their experiment further from the conditions required to yield a true Fickian permeability coefficient but which are more realistic in the context of occupational or environmental exposures and may identify more relevant rate-limiting phenomena associated with "real-world" systems, such as changes in solubility due to volatilisation of solvents or skin binding, which significantly influence transient or finite-dose exposures. Riviere and Brooks also comment that such modifications to the somewhat theoretical abstraction of Potts and Guy-type models—specifically by consideration of formulation effects—extend the relevance of their model beyond the specific experimental protocols they employ. This still, however, suggests that their model is only relevant to such systems in a broad sense (i.e. those that deal with transient exposures and volatility of solvents in the applied formulation) and extrapolations from other studies which employ different methodologies, particularly infinite-dose studies where steady-state diffusion occurs, may not be appropriate. In this context, they also suggest that their approach does not sit comfortably with the solvatochromic approach as the solvent mixtures used in their study may alter interactions in solution. Thus, in comparison with the Abraham and Martin approach, they find that its application to their range of solvent mixtures results in a series of QSPRs in which different descriptors are considered relevant compared to the original model. They suggest that this modification results in a model which again has more "real-world" relevance and that the relevance of different solvent mixtures is demonstrated in how they alter the rate-limiting processes of skin absorption, implying that vehicle effects on skin are more significant than solubility effects in solution in understanding skin permeation.

Finally, a recent study by Guth et al. (2014) has developed an in silico model for the estimation of absorption of chemicals from complex agrochemical formulations, based on previous studies by Riviere and colleagues which explored mixture effects in topically applied formulations. As before, they used a range of penetrants in a variety of formulations (in this case, 56 chemicals applied in more than 150 mixtures), obtained from regulatory in vitro studies with human and rat skin. They generated a wide range of physicochemical descriptors, and using multiple linear regression methods, they developed a statistically robust model ($r = 0.75$) which provides an opportunity to develop a relevant model that is able to estimate the influence of a new formulation on the dermal absorption of a well-known substance.

Conclusions

This chapter provides a brief overview of some of the studies where transient application of topical formulations, or the application of formulations with volatile ingredients, has been explored and shows attempts to model and quantify this. In some respects, it is difficult to decouple considerations of formulation effects in such systems and the approaches outlined above have attempted to consider a very difficult concept in—and address a significant limitation of—modelling skin absorption. Such work is still a small element of this wider field and is relatively new, and while our understanding of such systems reflects this, it is clear that this is an important field that has significant implications for risk assessment and dermal deposition and delivery.

References

Abraham MH, Martins F (2004) Human skin permeation and partition: general linear free-energy relationship analyses. J Pharm Sci 93:1508–1523

Anissimov YG, Roberts MS (1999) Diffusion modelling of percutaneous absorption kinetics. 1. Effects of flow rate, receptor sampling rate, and viable epidermal resistance for a constant donor concentration. J Pharm Sci 88: 1201–1209

Anissimov YG, Roberts MS (2001) Diffusion modelling of percutaneous absorption kinetics. 2. Finite vehicle volume and solvent deposited solids. J Pharm Sci 90:504–520

Baynes RE, Brooks JD, Budsaba K, Smith CE, Riviere JE (2001) Mixture effects of JP-8 additives on the dermal disposition of jet fuel components. Tox Appl Pharmacol 175:269–281

Baynes RE, Brooks JD, Riviere JE (2000) Membrane transport of naphthalene and dodecane in jet fuel. Tox Ind Health 16:225–238

Bronaugh RL, Stewart RF (1984) Methods for in vitro percutaneous absorption studies III. Hydrophobic compounds. J Pharm Sci 73:1255–1258

Buist HE, van Burgsteden JA, Freidig AP, Maas WJ, van de Sandt JJ (2010) New in vitro dermal absorption database and the prediction of dermal absorption under finite conditions for risk assessment. Regul Toxicol Pharmacol 57:200–209

Bunge AL, Cleek RL (1995) A new method for estimating dermal absorption from chemical exposure. 2. Effect of molecular weight and octanol-water partitioning. Pharm Res 12:88–95

Cilurzo F, Alberti E, Minghetti P, Gennari CGM, Casiraghi A, Montanari L (2010) Effect of drug chirality on the skin permeability of ibuprofen. Int J Pharm 386:71–76

Cleek RL, Bunge AL (1993) A new method for estimating dermal absorption from chemical exposure. 1. General approach. Pharm Res 10:497–506

Crank J (1975) The mathematics of diffusion, 2nd edn. Clarendon Press, Oxford

Cross SE, Anissimov YG, Magnusson BM, Roberts MS (2003) Bovine serum albumin containing receptor phase better predicts transdermal absorption parameters for lipophilic compounds. J Invest Dermatol 120:589–591

Dancik Y, Miller MA, Jaworska J, Kasting GB (2013) Design and performance of a spreadsheet-based model for estimating bioavailability of chemicals from dermal exposure. Adv Drug Del Rev 65:221–236

Diudea MV, Gutman I (1998) Wiener-type topological indices. Croat Chem Acta 71:21–51

El Tayar N, Tsai RS, Testa B, Carrupt PA, Hansch C, Leo A (1991) Percutaneous penetration of drugs—a quantitative structure-permeability releationship study. J Pharm Sci 80:744–749

EU (2004) Guidance document on dermal absorption. Sanco/222/2000 rev.7. Brussels, European Commission, Health and Consumer Protection Directorate-General. Directorate E—Food Safety: plant health, animal health and welfare, international questions, E1—Plant health. Available at http://ec.europa.eu/food/plant/protection/evaluation/guidance/wrkdoc20_rev_en.pdf

Flynn GL (1990) Physicochemical determinants of skin absorption. In: Gerrity TR, Henry CJ (eds) Principles of route-to-route extrapolation for risk assessment. Elsevier, New York, pp 93–127

Franz TJ, Lehman PA, Franz SF, Demetrulias JL, North-Root H, Kelling CK, Moloney SJ, Gettings SD (1993) Percutaneous penetration of *N*-nitrosoethanolamine though human skin (in vitro); comparison of finite and infinite dose applications from cosmetic vehicles. Fundam Appl Toxicol 21:213–221

Frasch HF, Barbero AM (2008) The transient dermal exposure: theory and experimental examples using skin and silicone membranes. J Pharm Sci 97:1578–1592

Ghafourian T, Samaras EG, Brooks JD, Riviere J (2010a) Modelling the effect of mixture components on permeation through skin. Int J Pharm 398:28–32

Ghafourian T, Samaras EG, Brooks JD, Riviere JE (2010b) Validated models for predicting skin penetration from different vehicles. Eur J Pharm Sci 41:612–616

Guth K, Riviere JE, Brooks JD, Dammann M, Fabian E, van Ravenzwaay B, Schafer-Korting M, Landsiedel R (2014) In silico models to predict dermal absorption from complex agrochemical formulations. SAR QSAR Env Res 25:565–588

Hansch C, Clayton JM (1973) Lipophilic character and biological-activity of drugs. 2. Parabolic case. J Pharm Sci 62:1–21

Kasting GB, Miller MA (2006) Kinetics of finite dose absorption through skin 2: volatile compounds. J Pharm Sci 95:268–280

Kruse J, Golden D, Wilkinson S, Williams F, Kezic S, Corish J (2007) Analysis, interpretation and extrapolation of dermal permeation data using diffusion-based mathematical models. J Pharm Sci 96:682–703

Kubinyi H (1977) Quantitative structure–activity-relationships. 7. Bilinear model, a new model for nonlinear dependence of biological-activity on hydrophobic character. J Med Chem 20:625–629

Lee CK, Uchida T, Kitagawa K, Yagi A, Kim NS, Goto S (1994) Relationship between lipophilicity and skin permeability of various drugs from an ethanol/water/lauric acid system. Biol Pharm Bull 17:1421–1424

Miller MA, Bhatt V, Kasting GB (2005) Dose and airflow dependence of benzyl alcohol disposition on skin. J Pharm Sci 95:281–291

Patel H, ten Berge W, Cronin MTD (2002) Quantitative structure-activity relationships (QSARs) for prediction of skin permeation of exogenous chemicals. Chemosphere 48:603–613

Pedretti A, Villa L, Vistoli G (2002) Modeling of binding modes and inhibition mechanism of some natural ligands of farnesyl transferase using molecular docking. J Med Chem 45:1460–1465

Potts RO, Guy RH (1992) Predicting skin permeability. Pharm Res 9:663–669

Riviere JE, Brooks JD, Monteiro-Riviere NA, Badsaba K, Smith CE (1999) Dermal absorption and distribution of topically dosed jet fuels Jet-A, JP-8, and JP-8(100). Tox Appl Pharmacol 160:60–75

Riviere JE, Brooks JD, Qiao GL (2000) Methods of assessing the percutaneous absorption of volatile chemicals in isolated perfused skin: studies with chloropentafluorobenzene and dichlorobenzene. Tox Meth 10:265–281

Riviere JE, Brooks JD (2005) Predicting skin permeability from complex chemical mixtures. Toxicol Appl Pharmacol 208:99–100

Riviere JE, Brooks JD (2011) Predicting skin permeability from complex chemical mixtures: dependency of quantitative structure permeation relationships on biology of skin models used. Tox Sci 119:224–232

Riviere JE, Brooks JD (2007) Prediction of dermal absorption from complex chemical mixtures: incorporation of vehicle effects and interactions into a QSAR framework. SAR QSAR Environ Res 18:31–44

Riviere JE, Qiao GL, Baynes RE, Brooks JD, Mumtaz M (2001) Mixture component effects on the in vitro dermal absorption of pentachlorophenol. Arch Tox 75:329–334

Saiyasombati P, Kasting GB (2003) Two-stage kinetic analysis of fragrance evaporation and absorption from skin. Int J Cos Sci 25:235–243

Slovokhotov YL, Batsanov AS, Howard JAK (2007) Molecular van der Waals symmetry affecting bulk properties of condensed phases: melting and boiling points. Struct Chem 18:477–491

Williams AC (2003) Transdermal and topical drug delivery. The Pharmaceutical Press, London

Wilschut A, ten Berge WF, Robinson PJ, McKone TE (1995) Estimating skin permeation: the validation of five mathematical models. Chemosphere 30:1275–1296

Chapter 9
The Devil is in the Detail...

I could write pages about the scene without giving a true idea of it.

George Orwell,
Down and Out in London and Paris, 1933.

Introduction

This book has reviewed the main research in the field of predictive modelling of percutaneous absorption. In doing so, it is inevitable that it cannot be exhaustive and that it will reflect the author's experiences and perceptions of the various contributions made to this field. Thus, in such a context, it is also important to address the underlying theme of this book: How has this field developed in the twenty-five years since the publication of Flynn's seminal study and in the twenty-three years since Potts and Guy's first—and most important—algorithm was published? This chapter will reflect on the generation of data from source through to the methods used—at times it will do so by following the somewhat cynical but often appropriate mantra of "garbage in, garbage out" from the field of computing, which really comments on how the inputs used to construct a model will influence the quality and significance of the outputs—and will reflect on both the foundations of the models and their robustness as well as their use, possible misuse or misinterpretation and their impact on the wider field of percutaneous absorption. In doing so, this chapter will necessarily critique the contributions discussed in the previous chapters in the context of impact and scientific relevance to the field.

Experimental Factors in Model Quality

Underpinning any model is the research work carried out, usually by other researchers, in generating in vitro permeability data. The need for accurate predictions of in vitro skin permeation is, as discussed widely in the earlier chapters,

G.P. Moss et al., *Predictive Methods in Percutaneous Absorption*,
DOI 10.1007/978-3-662-47371-9_9

highly significant to a number of fields, including pharmaceutical, cosmetic and toxicological applications. Therefore, the methodology for measurement of skin absorption in vitro has been widely considered as being reasonably standardised, providing comparable measurements of skin absorption rates for a wide range of skin penetrants (Bronaugh and Maibach 1985; Franz 1975; Grissom et al. 1987; Hotchkiss et al. 1990; Scott et al. 1992; Clowes et al. 1993; Reifenrath et al. 1994; Wagner et al. 2000; van de Sandt et al. 2002). Skin permeation is also associated with large variation due to the inherently diverse nature of the skin tissue often used for in vitro experiments, which may be exacerbated by particular protocols, such as the preparation, storage and pretreatment of skin tissue prior to experimentation (Michaels et al. 1975; Southwell et al. 1984; Kasting et al. 1994; Williams et al. 1992).

Thus, as well as considering the contents of this chapter in the context of Chap. 2, an important point of further context, when considering the quality of experimental data used to develop models of skin absorption, is the multicentre study reported by Chilcott et al. (2005). This was a study based across a number of invited laboratories which were each supplied with the same polydimethylsiloxane membrane and buffering chemicals (i.e. phosphate-buffered saline) and methyl parabens (the model permeant). Other chemicals were sourced locally by each participating laboratory and purchased separately. Instructions on the preparation of all solutions used in the study were also provided by the study coordinator. A "minimally prescriptive protocol", designed to provide a minimum set of rules from which standardised data could be obtained, was used by each participant, and the resultant data were analysed by the study coordinator to produce a "standardised" flux value by using the recommended analytical procedures specified in the study. The experimental protocol involved soaking the silicone membrane in water for 24 h prior to the experiment, using an absolute minimum of five replicates of the experiment, allowing 2-h equilibration time prior to the beginning of the experiment and applying the donor phase solution (1 mL of saturated methyl paraben solution, "ensuring that undissolved methyl parabens (MP) is present in the solution"). The donor compartment was occluded, and samples were collected regularly throughout a 24-h period. The effects of cell type, flow rate, receptor chamber volume, sample volume, and diffusion surface area on standardised fluxes were evaluated statistically.

Despite the thorough protocol employed, an average of 35 % variation was reported in fluxes of methyl parabens between the eighteen participating laboratories. This represented a four- to sixfold difference between the minimum and maximum fluxes for methyl parabens. Thus, while skin has been shown to be inherently variable (Michaels et al. 1975; Southwell et al. 1984; Kasting et al. 1994; Williams et al. 1992), removing this variable from Chilcott's study did little to increase the consistency of data produced across the 18 participating laboratories. Nevertheless, their analysis allowed them to conclude that receptor chamber volume, surface area available for diffusion, receptor sample volume and the design of diffusion cell were not significant factors in influencing the flux measured. Rather, they attributed the error found across the study participants to human error and experimental factors, including the membrane temperature (and not experimental receptor compartment temperature). They also recommended that a single analytical instrument based at a single laboratory might help in ensuring that the data obtained were more consistent. Finally, while they also commented that a similar study should be conducted with skin tissue, no such study has, at the time of writing, been published.

The implications of Chilcott's study are highly significant for the development of predictive models that rely on such experimental data. Clearly, if the inherent variation in a study where an artificial membrane—widely used in the skin permeation field as a homogenous alternative to heterogeneous skin tissue—is used is as high as 35 %, then it will be difficult for any algorithm (or similar mathematical method) to provide consistent predictions to any high degree of accuracy, particularly when membranes which are inherently variable, such as mammalian skin, are used. This variation in permeation across a consistent, homogeneous membrane might reflect the often statistically poor quality of some published QSAR models of skin absorption as the data, particularly in studies where small data sets are used to derive an algorithm, may simply be too inconsistent to produce a meaningful output, reflecting a poor predictive ability for such a model. Further, the mechanistic inferences drawn from even the best of models might be questioned, particularly when the significance, or lack of significance, attributed to particular discrete physicochemical parameters may be based on highly variable data. If, for example, a particular parameters' significance or non-significance is dependent on fine statistical margins, then it is reasonable to assume that variance in the source (input) data might be responsible for producing different and seemingly contradictory models, such as those discussed in Chap. 4. Such considerations also have implications for methods which report highly significant models, often with very small data sets, and whose statistical accuracy approaches unity (where, conversely, the issue of perceived operator variance may even be absent).

Thus, with such variable foundations in data consistency, the very nature of many such models can be called into question. It might also highlight a significant notational omission common to many models—while some researchers (including, e.g., Abrahams et al. 1995, 1997, 1999; Baert et al. 2007) commonly report their output algorithms as mean values of each significant parameter *and also report the standard deviation*, this is by no means a common practice in the presentation of algorithm-based models of percutaneous absorption. It is a practice which would provide additional information on the variability associated with the model and which would make clear the variability associated with a prediction as, in general, predictions from models are considered only as a single number which is in fact the mean prediction. An example of this might be that a researcher, unfamiliar with such practices, would use an algorithm by entering single numbers for each relevant physicochemical parameter, producing a single prediction. By including the variance of the output model into the prediction, a *range*, rather than a single number, would be output—this would provide the user of the model with a more biologically realistic model for his or her studies and place the use of the model in a more appropriate context. Thus, the output from such models should be presented as suggested by Chilcott et al. (2005) as the μ value (which, in their study, represented the 95 % confidence interval). An example of where such variance has been openly described was presented by Sun et al. (2011).

Moss and Cronin (2002) addressed the issue of data quality by reanalysing the Flynn (1990) data set. Their study was informed by the work of Johnson et al. (1995) who reported spurious steroid permeability data in the literature. In particular, they reported that the data which formed part of Flynn's data set, which was originally published by Scheuplein et al. (1969), differed significantly from other, similar data in the literature. This is summarised in Table 4.1, where it can be seen that the steroid permeability data published by Scheuplein et al. differ significantly from the data published by a number of other researchers. Thus, they concluded that this data subset, which is a significant part of the larger data set (approximately 15 %), contained erroneous data. Moss and Cronin reanalysed these data and obtained a QSAR (shown in Chap. 4 as Eq. 4.30) which is very similar to the Potts and Guy's (1992) algorithm. The Moss and Cronin's algorithm differs from the Potts and Guy's equation as the coefficients for the significant terms are weighted differently. However, while this might suggest similar predictive performance, Moss et al. (2006) demonstrated that the Moss and Cronin's equation did not overestimate flux as much as the Potts and Guy's equation. This is summarised in Fig. 9.1. Further, it was generally considered, following the Potts and Guy's (1992) article, that the steroids did not fit the model and therefore should be considered as outliers; this also inferred that steroids permeated the skin by a different mechanism than the other chemicals in the data set, a phenomenon attributed to the number of potential hydrogen bonding sites present on such permeants. Thus, it is clear that the steroid data in Flynn's data set should be replaced by the data collated by Johnson et al. and used by Moss and Cronin in developing their QSAR. It also highlights the pitfalls of collecting literature data and adding it to a data set. Unfortunately, it may be impossible to validate all such data, given the nature of data variability raised by Chilcott et al. (2005) and therefore the value of models thus derived. This also, logically, suggests that if erroneous data are included in a model, its impact—which is normally attributed to the biological variability inherent in the tissue or operator issues—may mask such findings. Clearly, this has been the case with the Scheuplein data used by Flynn, Potts and Guy, and others. For example, in comparing the Potts and Guy's (1992) and Moss and Cronin's (2002) algorithms with new experimental data, Moss et al. (2006) found that despite superficial similarities, significantly different predictions were obtained from both algorithms, neither of which, incidentally, matched in magnitude or trend the experimental results reported.

It is also striking that despite approximately 15 % of its data having been shown to be erroneous, the Potts and Guy (1992) algorithm—and others based on this study—is still used by researchers and other models, which incorporate the corrected data, find little utility. Whether this is through a sense of the real strength of Potts and Guy's model—its simplicity and ease of use—in the context of an inherently variable data environment, or a sense that the points raised by Johnson et al. (1995) do not significantly alter its key message (that permeability is

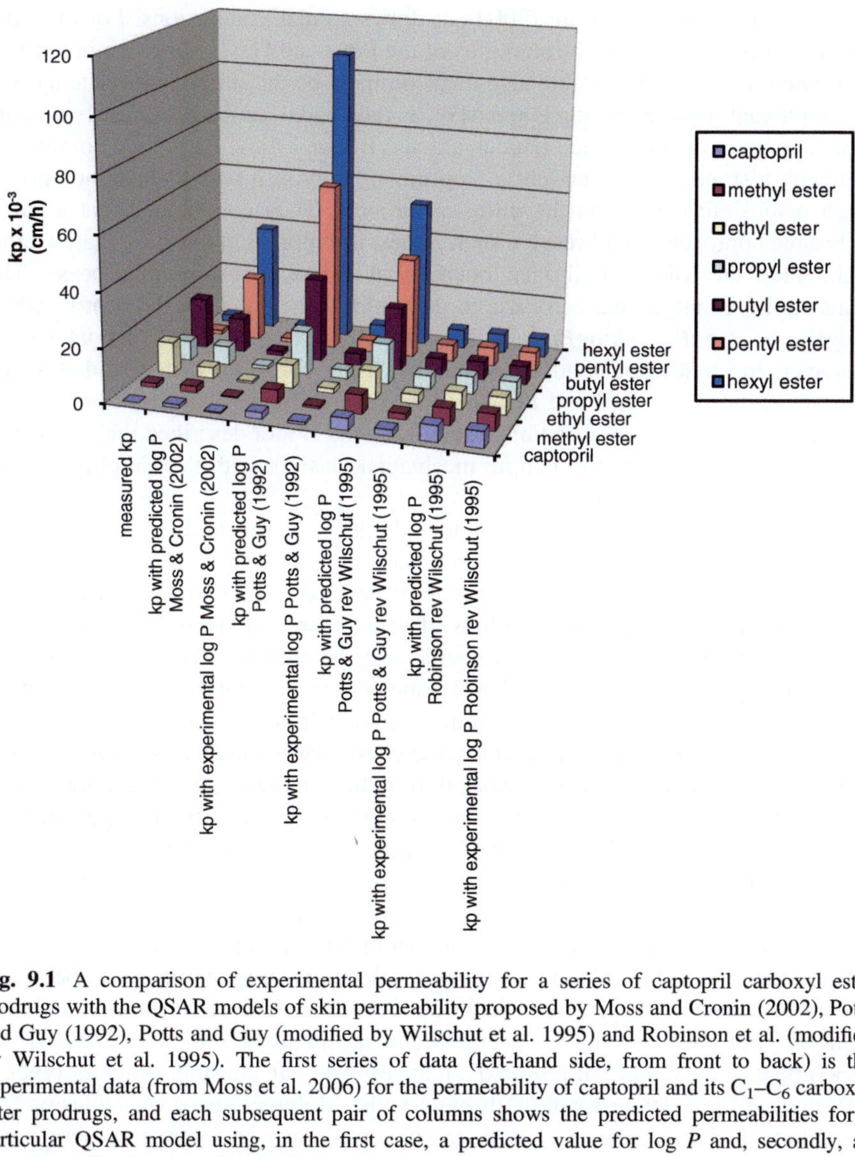

Fig. 9.1 A comparison of experimental permeability for a series of captopril carboxyl ester prodrugs with the QSAR models of skin permeability proposed by Moss and Cronin (2002), Potts and Guy (1992), Potts and Guy (modified by Wilschut et al. 1995) and Robinson et al. (modified by Wilschut et al. 1995). The first series of data (left-hand side, from front to back) is the experimental data (from Moss et al. 2006) for the permeability of captopril and its C_1–C_6 carboxyl ester prodrugs, and each subsequent pair of columns shows the predicted permeabilities for a particular QSAR model using, in the first case, a predicted value for log *P* and, secondly, an experimentally determined value for log *P*

dependent upon lipophilicity and molecular weight) is unclear. In this latter context, it should also be commented that when these erroneous data were corrected and reanalysed, the outcome was very similar to that originally presented by Potts and Guy (Moss and Cronin 2002).

Outcomes similar to those reported by Moss and Cronin (2002) were also found by Buchwald and Bodor (2001). While their study cannot be directly compared to the work of Moss and Cronin (2002), it draws similar conclusions. For example, they not only comment on the strengths of the Potts and Guy's approach in terms of transparency and simplicity, but also then comment on the strong interrelatedness of the significant parameters, log P and MW, as they had described previously how the former is strongly size-related (Buchwald and Bodor 1998). Thus, they commented that such a strategy is effectively a "top-down" approach which sheds no specific mechanistic information on the diffusion process. They commented that a solvatochromic approach would resolve such issues; as reported in Chap. 4, these studies highlighted the role of hydrogen bonding in the skin absorption process. This includes, for example, the early use in this field of the $\Delta\log P_{o/h}$ descriptor (where $\Delta\log P_{o/h} = \Delta\log P_{o/w} - \Delta\log P_{heptane/w}$) by El Tayar et al. (1991) and its role in characterising hydrogen bond donor acidity. They commented that studies where data are collated together and not analysed separately in subgroups appear to be contradictory to El Tayar's findings. Buchwald and Bodor described the application of their "QLogP" method to couple mechanistic insight with the simplicity of the Potts and Guy's approach.

Sitting above the attempts of Chilcott et al. (2005) to standardise methodology in the measurement of in vitro percutaneous absorption is a diverse range of experimental methods from which such data can be derived—all of which may contribute to data sets used to develop algorithms of skin absorption. Thus, it is important to remember that the experimental protocols may be significantly different and that they reflect primarily the needs of experimenters in particular studies and not the needs of modellers who will later abstract such data from those studies for their own purposes. In a particular in vitro study, there are a large number of parameters that can be altered, by design or otherwise, during the conduct of the experiment. Thus, in the collation and use of such data for the development of structure–permeability relationships, there is the possibility of data being compiled from experiments carried out by a wide range of methods. This may affect the relevance of the models subsequently developed and the validity of their output.

The relevance and advantages of conducting in vitro studies, particularly in the earlier stages of a formulation or drug development programme, have been discussed in detail in Chap. 2. The main benefits of such tests are in cost and convenience, but they also allow specific technical modifications to be made to the experiment. For example, if required, an experiment can consider different types of diffusion cell (i.e. static for flow-through cells), different states of occlusion, the condition of skin (where, e.g., a researcher may have an interest in permeation following chemical or physical depilation), the use of different solvents in either the donor or receptor compartments of the diffusion cell and a range of experimental temperatures. Such variations in experimental technique can allow significant control of the laboratory environment and therefore specific phenomena to be studied, such as the excellent thermodynamic studies conducted by Hadgraft and colleagues, discussed in Chap. 5. They also allow important protocols for the understanding of toxicity and permeation to be developed, where the laboratory advantages of radiochemical labelling techniques can be exploited in a manner that is not appropriate in in vivo studies. Clearly, they do not consider issues such as blood flow and metabolism, and the constraints and limitations of any in vitro method, when compared to in vivo testing in particular, should be understood and any results treated in the appropriate context. It thus follows that the same caveats should apply to mathematical algorithms derived from such data, as well as being mechanistically understood in the context discussed by Buchwald and Bodor (2001).

Further issues that may introduce possibly significant variance into experimental protocols include the selection and treatment of skin. Standardised measures of skin integrity (such as measurement of transepidermal water loss, the pretreatment with ^3H or measurement of electrical conductance) will allow the skin barrier to be characterised and for particular samples to be discarded if they fall outside acceptable ranges of use. In addition, individual researchers may prepare skin samples differently depending on specific needs or to adhere, for example, to protocols used in previous similar studies. This will include preparing the skin by removing all or part of the tissue underlying the epidermis (the dermis and/or the subcutaneous fat), which is normally achieved by the application of heat, with the use of enzymes or mechanically, for example, with a dermatome. Whether the skin used for a permeation study is used fresh or used following freezing and storage may also impact on the permeability measured.

Chapter 2 introduced the concept of diffusion cells for in vitro measurement of permeation across biological membranes, including skin. Most cells are, essentially, variations on a theme, with the majority of diffusion cells currently in use being mostly based on designs by Franz (1975). Thus, they will have two compartments, and the permeant's ability to pass from one compartment (the donor) to the other (the receptor, which is normally stirred to ensure uniformity) via a membrane (such as mammalian skin or an artificial membrane), is determined. Cells are normally arranged vertically or side by side (where both compartments are usually stirred). Stirring is essential in order to avoid or minimise the occurrence of static diffusion layers within the diffusion cell and to avoid high local concentrations of permeant which may influence the diffusive process, affecting the results of a study (Stehle and Higuchi 1972; Lovering and Black 1974). The compartments will usually be maintained at a specific temperature which is predominately, but not exclusively, 37 °C ($\pm$1 °C) in the receptor compartment which equates to 32 °C ($\pm$1 °C) at the skin surface (Barry 1983). Inefficiencies in temperature control, even to a very small degree, can significantly influence the permeability measured in an experiment (Friend 1992; Chilcott et al. 2005), leading to a substantial source of error if such data are subsequently collated and used for model development.

The nature of the quantitative analytical method may also influence, albeit minimally, accuracy of measurements in skin permeability studies. Methods commonly used include chromatography, spectroscopy, scintillation counting and specific assays which may be biological in nature (Nugent and Wood 1980). For non-radiolabelled techniques applied to, for example, cosmetic systems, a lack of specificity in penetrants (such as triglyceride-containing natural oils or proteins which are found in some topically applied cosmetic products) may be an issue, particularly when they are derived from natural sources which may contain a range of materials which may not necessarily all be at the same concentration in all samples and which may also be similar to material found naturally within the skin. While more recent analytical techniques, particularly those incorporating mass spectrometry, may reduce or eliminate such considerations, the context of collating data from the literature may include doing so from experiments where such analytical issues were not, or could not be, addressed. Thus, care in collation of the data set for development of a mathematical model of skin absorption should extend to an understanding of the analytical processes used in the original studies from which the data were collated.

Where extraction of a material from a biological matrix forms the part of an analytical procedure, this may result in incomplete or variable recovery. One generalisation often made in the collation of skin permeability data appears to be the potential for experimental variation, particularly when conducting bioanalytical studies; factors such as stability, storage, inter- and intra-day variation are seldom accounted for, and the general assumption has been that samples can be measured properly and consistently, when in fact such analyses are prone to variance. Ideally, it would be for the benefit of models produced for the data they use to be taken from guidelines, such as the FDA guidelines for bioanalytical method development. This may extend to consideration of the nature of the solvents used in both donor and receptor compartments, solute solubility in those compartments and the effects they may exert on the skin barrier function (i.e. in the case of solvents which may, for example, delipidise the skin) or on solubility of potential permeants in the receptor compartment.

Composition of the receptor phase can thus influence solute solubility and therefore may affect the nature of the permeation process being investigated (Barry 1983; Bronaugh and Maibach 1999; Williams 2003). It is common, in measurements of in vitro percutaneous absorption, to modify the receptor composition in order to facilitate diffusion of a permeant into this compartment. It should be noted that the general models from Flynn onward are based on infinite-dose experiments also consider diffusion from saturated aqueous solutions only, as Flynn commented that it is from such vehicles that the bulk of environmental exposures would occur. Thus, in vitro systems where the receptor compartment contains a solvent other than water (or a simple buffer) are not normally considered for inclusion in data sets from which mathematical algorithms are produced.

The composition of the receptor phase will therefore be highly significant in influencing percutaneous absorption, not just by its volume and mixing but by its chemical composition. For example, Kasting et al. (1987) used a receptor phase solution which contained 50 % v/v ethanol as well as a small amount of sodium azide (0.02 % w/w) as an antimicrobial agent. This receptor phase was chosen as it would ensure solubility of their penetrants in this compartment. Bovine serum albumin (BSA) has been added to receptor compartments, in concentrations as high as 4 % w/w, by many researchers, including Dal Pozzo et al. (1991) and Sartorelli et al. (1998) as it assists in the solubilisation of penetrants with a wide range of lipophilicities. Lin et al. (1996) used a receptor phase buffered at pH 6 to ensure the solubility of their permeant, which is pH dependant, throughout their experiment, and Mistry et al. (2013) employed a citrate buffer to ensure that their permeant, aluminium, was solubilised throughout their experiment as this chemical forms an insoluble oxide at neutral pH.

Such modifications to experimental protocols clearly move them further from the in vivo ideal, making in vitro–in vivo correlations more difficult to quantify, but also clearly signposts that these experiments are being carried out in vitro and should be understood in this context. A range of other factors should be considered in the context of developing in vitro models of skin absorption. These include the effect of the systemic circulation and biological processes such as active transport or metabolism. Somewhat idealistically and impracticably, it may be proposed that the most representative receptor fluid may be either human or animal blood (or plasma), although it may have to be modified to include anticoagulants and its use may raise particular handling concerns (Moss et al. 2012).

Therefore, despite commonality in generic experimental design, and attempts such as those by Chilcott et al. (2005) to develop common approaches, little standardised methodology has been adopted for the measurement of percutaneous absorption. Given the main reason for conducting such studies is not often related to developing mathematical models of permeation, this limits significantly the amount of data available in the literature from which models can be developed. It also limits the usefulness and scope of models so developed as they may be of little relevance to other systems.

Considerations relating to finite-dose experiments have been discussed previously in Chap. 8.

Analysis of the Experiments from Which Data Have Been Taken to Develop Models of Skin Absorption

The preceding section outlines the range of experimental factors that can be used in a seemingly simple Franz-type diffusion cell-based experiment. The excellent review by Friend (1992) describes in detail the range of experimental techniques available for such seemingly straightforward measurements. This experimental variance is something which underpins the data used to develop quantitative models of skin permeation and highlights the need to examine the source data used. This is again underpinned by the Scheuplein data added to the Flynn data set, which varied considerably from findings from a number of other researchers (Michaels et al. 1975; Hadgraft and Ridout 1987; Goodman and Barry 1988; Hou and Flynn 1989; Liu et al. 1991; Williams et al. 1992; Knutson et al. 1993; Yum et al. 1994; Mitragotri et al. 1995). Of the fourteen steroids examined by Scheuplein, six (aldosterone, corticosterone, estradiol, hydrocortisone, progesterone and testosterone) have measured permeabilities lower than those reported by other researchers by factors of between 5 and 77. For eight steroids, the published permeability data are all in good agreement with each other but vary from the Scheuplein data by factors of between 11 and 20. It is also interesting to note that in Johnson's evaluation of these data, they considered potential experimental factors, such as those outlined in the previous section, but concluded that such factors as experimental temperature, establishment of steady-state conditions, methods of skin preparation and the use of radiolabelled permeants did not influence the studies and that the data are incorrect.

One point often not highlighted in the commentary of the Scheuplein data is that while other researcher's permeability data are in agreement, this essentially means that it is within one order of magnitude. This is still a significant margin of error and indicates that even controlled studies which offer a broad consensus do so within the wider context of methodological and membrane variability which is difficult to control or minimise, despite substantial efforts in trying to harmonise methodology, including skin barrier preparation. It is also reflected in the statistical approach taken in preparing data for analysis. In general, it might be considered that all data —unless reasonably determined to be unsuitable—should be included for analysis. Whether this is then examined as a mean or median value of "repeated" data (i.e. permeability data for the same chemical from different experiments) or whether each individual data point is used in constructing a data set is an important point to be considered, as it is often a point that is irregularly presented with sufficient transparency—if at all—in some studies.

It is also insufficient to simply collate different values and then use the resultant mean or median. For example, in the steroid data summarised by Johnson et al. (1995), they showed that the different experiments summarised in Table 4.1 were conducted at different temperatures of 25, 26, 30, 32 and 37 °C. It was argued by Johnson et al. that these data could be temperature-corrected using the Arrhenius equation, resulting in only a twofold change in skin permeability. While such corrections clearly often improve the consistency of data, they still demonstrate a significant variance in the data and, it might be argued, a lack of clarity in presenting this point in their study so as to avoid the use of data obtained at different experimental temperatures without further adjustment. More broadly, it is often unclear from published studies whether the temperature quoted relates to the membrane temperature (i.e. the skin surface temperature) or the temperature in the receptor compartment of the diffusion cell.

The points raised above have resulted in some researchers challenging the very concept of developing mathematical models of skin absorption from experimental data. Not only are such models limited by the formulation type applied to the donor compartment, they are beset also by variable data which are difficult to validate (e.g. Scheuplein's original, disputed data have a variance of approximately 25 %, which generally meant that skin researchers felt that it was reasonable given the inherently variable nature of the barrier being used), they may not properly address the variability in their output, and they are prone to misinterpretation and misuse.

For example, Walters and Brain (2000) commented on the lack of relevance of QSAR-type models of skin absorption, which followed the comments outlined above. In particular, they highlighted where such models were not relevant by considering a range of permeant/vehicle interactions and choices of solvents used in experiments from which models are developed. However, rather than commenting on the lack of relevance of such models, it may be argued that they addressed the issue of model perception and misuse, highlighting the limitations of a range of models rather than their redundancy. As an example, they commented that as most Potts and Guy-type models for skin absorption relate to permeation from aqueous solutions, any predictions or mechanistic insight would be relevant for such systems only. This is indeed the case, and both the reasons for the use of such a system and its potential limitations were highlighted in Flynn's original work. What Walters and Brain do not offer is a solution to these limitations, but this is an issue addressed by others, which will be discussed below. So, while they cited earlier work of theirs (Walters et al. 1997) which highlighted the differences between formulations and how this has a significant effect on permeation and flux, this essentially highlights that modelling of percutaneous absorption is still a relatively new field of research and that it is one which has led to significant insights to the wider issue of skin permeation—this is seen from the outcomes of many studies discussed in Chaps. 4–6 in particular and also in the formulation-focused studies discussed in Chap. 8. However, the misuse or misinterpretation of models is the real underlying issue in such studies. The mathematical algorithms developed for skin permeation are limited to their input data and while significant criticism is valid—for example, that which relates to data quality, variance in experimental results and membrane variance—other criticisms of models seem to be informed from an incomplete knowledge of model construction and validation, as well as a lack of understanding of their domains of applicability. Nevertheless, the issues raised by Walters and Brain are significant in the context of formulation and solvent effects.

Systematic studies examining models of skin absorption were conducted previously (Moss et al. 2005, 2006). They synthesised and characterised a series of new prodrugs of captopril, an ACE inhibitor whose potential as a transdermal therapeutic is based on the maintenance of a steady blood plasma concentration over a fixed period of time. They modified the parent drug at the thiol and carboxyl groups, which had a wide range of lipophilicities and molecular weights—all of which were still within the "molecular space" of the Flynn's data set. Predictions of permeability were obtained from a number of QSAR-type models and were compared to in vitro Franz-type diffusion cell studies using a variety of membranes, including fresh and frozen porcine skin and a polydimethylsiloxane membrane.

The findings of both studies are summarised in Fig. 9.1. This compares experimental permeabilities for a series of captopril carboxyl ester prodrugs with predictions from a number of QSAR-type algorithms. While the model is limited in the nature of the comparison—given the species used experimentally—it demonstrates several fundamental issues with the use of QSAR-type models of skin permeation. The first clear issue is that despite any limitations due to the membrane used, the trends for measured permeability in skin are, in most cases, significantly different from the predictions in a number of ways. They initially show reasonable agreement between experimental and predicted permeabilities for the less lipophilic prodrugs, generally up to a point where the log P of the permeant is at, or below, 2.0.

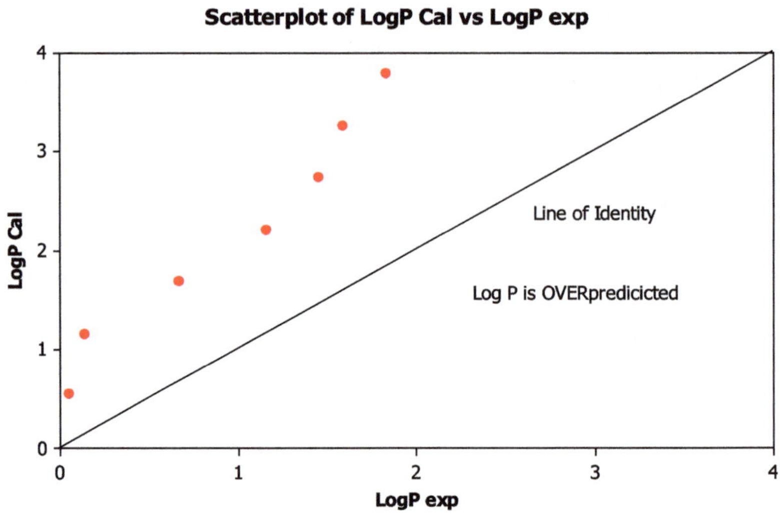

Fig. 9.2 A plot of experimentally measured log *P* against calculated log *P* for a series of captopril prodrugs, showing the deviation from linearity associated with increasing lipophilicity (modified from Moss et al. 2005)

However, above this point, the models begin to differ significantly from the experimental findings, which are a set of results that empirically agree with our understanding of skin permeability outside the context of algorithm development. Specifically, the experimental results show an optimum permeability (generally around the C_2–C_4 prodrugs), followed thereafter by a decrease; similar conclusions have been drawn in the earlier chapters in the context of nonlinear modelling. So, while the general trends between experimental and predicted permeability data do not compare across the stated range of validity for the prodrugs examined, another significant issue which is clearly apparent in Fig. 9.1 is the overestimation of permeability by most models compared to experimental findings. This may be due to a number of factors—for example, issues with skin thickness may influence such direct comparisons—but the scale of overestimation is significantly larger than any such experimental factors, which in themselves are not considered by the models as they are based on data from a range of diverse experiments. Overestimation is at its worst for the prodrugs with the highest lipophilicity.

Such overestimation may be explained, at least partly, by considering the effect that calculated log *P* values have on predictions from the models. In most cases shown in Fig. 9.1—the Robinson, revised by Wilschut, model being the obvious exception—the use of calculated log *P* values results in significant overprediction of k_p. Figure 9.2 (modified from Moss et al. 2005) compares experimentally determined log *P* values with calculated values (calculated values of log *P* were determined using a range of resources, including the IA program, available at http://www.logp.com, the EPI Suite software (version 3.12) and the Virtual Computational Chemistry

Laboratory resource, available at http://146.107.217.178/). Following the methods proposed by Ranadive et al. (1992), they used the predictions that were closest to their experimental measures. A significant deviation from the line of identity is observed as log *P* increases, suggesting that the predictions for log *P* are less accurate as lipophilicity increases. This may also explain the gross overprediction associated with predictions from some of the QSAR models shown in Fig. 9.1. In addition, their methodology—in selecting the prediction closest to their experimental results—does also suggest that worse predictions could increase the error associated with using such data in QSAR models of skin absorption. Also, Moss et al. (2006) reported that the melting points of their captopril carboxyl ester prodrugs did not agree with the predicted melting points (obtained from the EPI Suite software). The predictions suggested that all prodrugs would have higher melting points than the parent drug. However, when measured, all prodrugs had lower melting points and were, in fact, oils at room temperature, compared to the parent drug (captopril) which was a solid.

Thus, when the potential sources of error are considered in such models, they are often discussed in terms of the variability of experimental data and the underlying biological variation which is to be expected when using mammalian skin. However, such sources of inaccuracy are simply exacerbated by the limitations of predicting physicochemical descriptors. It also provides a clear issue both in using the models with such inputs and the implications that this will have for any outputs. It should also be considered that while predictions are improved when experimental values are used, the resultant predictions still bear little resemblance to experimental data over the *totality* of the molecular space considered to be valid based on the range of input data (i.e. Potts and Guy 1992; Mitrogotri et al. 2011). In addition, when the trends in predictions from the QSAR models are considered, it may be inferred that methods that fit linear outputs to such data may not be appropriate. It should also be noted that the outcomes described above, and the inferences drawn, are limited by the paucity of similar studies in the literature—further work is clearly required in order to develop a fuller understanding of such issues.

Therefore, the use of predicted/calculated values of physicochemical descriptors, such as log *P* and melting point, should be avoided if possible, or—at the very least—studies which use them should highlight which data are based on experimental findings or predictive software—while this is the case in some published data sets, it is not always the case, so predictions made from models, or even more significantly models developed from predicted descriptors which have been shown to compare poorly to measured values, should be questioned. This suggests that, where relevant, models should be developed from measured physicochemical descriptors and not from calculated parameters. It also raises the point addressed by Walters and Brain as to whether or not we should extend such considerations to the very use of QSAR-type models for prediction of skin permeability at all.

Such issues have also been raised by Geinoz et al. (2004), who critically evaluated QSAR models applied to skin permeation. They comprehensively reviewed the significance of a variety of models and their applicability. For example, they comment on the QSAR derived by Dearden et al. (2000), shown previously as Eq. 4.24. They comment on the use of a wide range of descriptors and, in contrast to comments such as those from Buchwald and Bodor (2001), criticise the number of descriptors used (81, even though not all were defined).

However, Dearden's study does raise one important limitation in other studies: it did not begin with the assumption that some descriptors very commonly used in the physical sciences (such as log *P*, MW, melting point and hydrogen bonding expressions) were those which *should* be used to model skin absorption. Rather, it examined as wide a range of descriptors as were available and appropriate to use, thus increasing the scope and validity of their work. While Geinoz et al. comment that the lack of data in Dearden's paper means that the evaluation of the relevance of these parameters is not clear, Dearden's approach is mechanistically transparent and highly specific; perhaps the issue is more to do with understanding and interpreting the relevant parameters rather than their fundamental nature. Geinoz does not comment on whether they are referring to the descriptors used by Dearden in general (all 81) or to only those which Dearden's analysis found to be statistically significant. Again, while physical scientists may work more with the generic (and often composite) descriptors discussed throughout this book, Dearden's approach is more focused on molecular structure and is more in keeping with QSAR methodologies found more broadly in this field, particularly outside the field of percutaneous absorption and which have been applied to this field by researchers such as Ghafourian and Fooladi (2001). Geinoz et al. do agree with Buchwald and Bodor's criticism of Potts and Guy's equation, in their comment that it may be difficult to evaluate the relevance of the terms used due to possible intercorrelation or covariance. Genioz et al. did also consider the model proposed by Buchwald and Bodor, dismissing it as potentially biased due to the nature of the data sets used, as more hydrogen bonding sites imply larger molecules, particularly for drug-like compounds. Finally, Geinoz et al. reflect on Moss and Cronin's interpretation of the data by Johnson et al. (1995) and comment that Flynn's original data set may be questioned. They also comment that a number of models lack statistical rigour and present apparently significant descriptors which are not relevant, thus overemphasising their role in providing specific mechanistic insight to the process of skin permeation. Their comments on the nature of Flynn's data set would infer therefore that a more complete data set on which to build new models is that presented by Moss and Cronin, which was used to develop a new QSAR using a wider range of physicochemical descriptors by Patel et al. (2002).

Sitting next to such considerations is the generic perception—which in some cases is seemingly supported by high statistical quality—that nonlinear models tend to provide better predictions of skin permeability, despite the myriad issues raised above. Lam et al. (2010) commented that certain machine learning methods had been prone to over-fitting data; this included artificial neural networks, which tend to over-fit when large numbers of predictors are employed as such models are often weighted and susceptible to overtraining (Neumann et al. 2006), resulting in idiosyncratic results, particularly as the output has a tendency to fit the noise associated with the data, resulting in a good statistical fit to a training set but often variable predictivity for new compounds (Guha and Jurs 2005). Hence, Lam et al. used Gaussian process methods as they tend to minimise, but not remove, such considerations. Further, it has been shown recently that the simple use of nonlinear methods does not automatically result in a model of improved statistical performance and better predictions for new compounds. Moss (2014) commented that in studies with a number of nonlinear machine learning methods, only marginal improvements in the statistical quality of models were observed despite increasing the power terms of models from n^2 to n^6, wherein it is the perception that such changes to the statistical analysis will result in a curve fitting as many data points as possible. Such observations, although related to specific systems and lacking in a wider context, do sit against the perception that nonlinear models will automatically provide better fits to data than linear models. Indeed, such perceptions sit in opposition to studies such as those by Hewitt et al. (2014), which found that a linear algorithm provided the best output following the analysis of a permeability data set for a polydimethylsiloxane membrane.

Formulation Factors

Following on from Chap. 8, one of the major limitations to the models of skin permeability discussed so far is their inability to model formulation effects. This, by and large, has been due to Flynn's observation that most environmental contaminants will be presented to the skin as aqueous solutions. Nevertheless, the nature of the formulation clearly plays a significant role in percutaneous absorption—numerous examples of how pharmaceutical drug delivery is enhanced by formulation manipulation are discussed by Williams (2003) in his excellent book. The solubility of a penetrant within the formulation including its exact form (and its bioavailability) which may be influenced by formulation pH, its interactions with the formulation components and the interaction of the formulation components—including the drug—with the skin barrier which, in some cases, has been shown to modify the barrier function, will all influence the ability of a penetrant to permeate into and across the skin. This is clearly a limitation with quantitative models of skin absorption that sits with the criticisms of such models, discussed above, but which has been addressed by a number of researchers. It is therefore important to use quantitative models, both mechanistically and predictively, only in their original context. The issue of solubility and formulation solvent selection is significant; as the most widely used term in modelling to describe skin permeation, the skin permeability coefficient (k_p) is essentially the flux across the skin normalised by solution concentrations. Thus, in different solvents, the fluxes and therefore permeabilities will differ significantly.

Several significant studies have investigated the incorporation of formulation effects into quantitative models of skin absorption. For example, Gahfourian et al. (2007) developed a QSPR model of drug release from a hydroxypropyl methylcellulose (HPMC) formulation—this is a widely used pharmaceutical agent, used in a wide range of solid and semi-solid formulations as a viscosity modifier or to modify the rate of release in oral modified-release dosage forms. A range of endpoints relevant to drug dissolution were employed, i.e. the time required for 50 % of the drug release to occur ($T_{50\ \%}$), percentage of drug released after 8 h ($Q_{8\ h}$), release rates (k) and dissolution efficiency (DE_8, the area under the dissolution curve at time t, where $t = 8$ h). They found that they were able to model these endpoints using a range of physicochemical descriptors for a data set of 14 pharmaceuticals and suggested that the aqueous solubility of the drug and its size were the most significant factors. They were able to demonstrate that drug release form the eroding matrix occurred via two routes, the balance of which depended on both the properties of the drug and the grade of polymer used in the formulation.

Such methods have been applied to the modelling of skin permeation. For example, Ghafourian et al. (2010), discussed in more detail in Chap. 8, explored quantitatively how vehicle effects in skin barrier function (and hence on diffusivity and permeability) can be modelled. They developed a QSAR to relate the skin permeability of 12 compounds, each prepared in 24 different solvent mixtures, in their data set to the chemical properties of the formulation mixture. They found that mixture composition could exert a significant influence on skin absorption. As importantly, they used different descriptors and endpoints to characterise their outcomes. For example, most QSAR models traditionally use either flux or the permeability coefficient and compare such outputs to the physicochemical properties of a data set of penetrants. However, in the case of work by Riviere and Ghafourian, they have used estimates of particular combinations of such descriptors. This is a powerful way to model complex phenomena such as formulation effects and offers a significantly different approach than measurement of direct influences of formulation, such as physical (i.e. rheological, wetting/swelling) changes in a dosage form or drug release profiles (i.e. use of T_{50} as an alternative endpoint to k_p). It is a similar approach to that considered by Moss et al. (2013), who examined a version of the data set published by Magnusson et al. (2004) but included descriptors other than physicochemical descriptions of a molecule. For example, they included experimentally relevant terms to their analysis and found that one such term, the experimental temperature, was returned as a significant descriptor:

$$\begin{gathered} \log J_{\max} = 0.071T_{\exp} - 0.01MW - 0.086MPt - 29 \\ [\text{Sum squared error} = 243.85 \quad r^2 = 0.90 \quad p < 0.001] \end{gathered} \tag{9.1}$$

where

T_{exp} is the experimental temperature reported by the authors of the original studies

MW is the molecular weight

MPt is the melting point

This result, in its inclusion of molecular weight and melting point as significant descriptors, is in general agreement with Magnusson's original study, but the addition of experimental temperature to the resulting QSAR improves the statistical quality of the original model.

Conclusions

This chapter has considered some of the underlying issues associated with the development of robust and valid models of skin absorption, and it has discussed some of the limitations associated with their construction, interpretation and use. In particular, emphasis is given to appreciating the whole process—from data generation and use through to interpretation and scope of relevance associated with a model. In combination with some of the research discussed in Chap. 8, it is clear that the models suffer from significant limitations—or a lack of understanding of their scope—and that misuse and misinterpretation are significant problems. Thus, while such models should provide mechanistic insight and predictive power, the former, in particular, is different—often substantially different—for each model developed and therefore questions the significance of the latter. Data have often been abstracted from larger data sets without robust rationale or with consideration of the possibility that such data are in itself comprised of outputs from different studies, whose potential variance is underlined by Chilcott's study. Thus, it may be the case that any data used to construct a model should be presented in such a way as to ensure that experimental conditions are described alongside the physicochemical descriptors, so that a true understanding of the output—and the effect of all the input variables—can be used to underpin consideration of any model derived. This is also emphasised by Geinoz et al. (2004) and Sun et al. (2011) who include in their description of QSPR models the mean and standard deviations associated with each descriptor, which provides a far more robust and practical description of the models they reviewed. It may also be the case that the inherent variance associated with some descriptors used to construct models, such as experimental or predicted values of lipophilicity (usually as log P or log $K_{o/w}$) or melting point, may themselves introduce error to the models and that the use of predictors not disposed to such variance might, in the context of algorithm-free models, such as Gaussian process regression, provide more accurate models.

Clearly, consistency in how models are developed and validated is required. The next chapter will discuss key approaches to developing robust and relevant models and will do so by looking at the wider context beyond studies of skin absorption. The underlying premise in the development of any such models is, in its most abrupt terms—whether they actually work in providing mechanistic insight to the processes of skin permeation and in providing accurate models of the skin absorption process and whether they can do so in the context of the underlying error associated with both the generation of experimental data (as described by Chilcott et al. 2005) and the inherent biological variation of skin.

References

Abraham MH, Chadha HS, Mitchell RC (1995) The factors that influence skin penetration of solutes. J Pharm Pharmacol 47:8–16
Abraham MH, Martins F, Mitchell RC (1997) Algorithms for skin permeability using hydrogen bond descriptors: the problem of steroids. J Pharm Pharmacol 49:858–865
Abraham MH, Chadha HS, Martins F, Mitchell RC, Bradbury MW, Gratton JA (1999) Hydrogen bonding part 46. A Review of the correlation and prediction of transport properties by an LFER method: physicochemical properties, brain penetration and skin permeability. Pest Sci 55:78–88
Baert B, Deconinck E, van Gele M, Slodicka M, Stoppie P, Bode S, Slegers G, van der Heyden Y, Lambert J, Beetens J, de Spiegeleer B (2007) Transdermal penetration behaviour of drugs: CART-clustering, QSPR and selection of model compounds. Bioorg Med Chem 15:6943–6955
Barry BW (1983) Dermatological formulations: percutaneous absorption. Marcel Dekker, New York
Bronaugh RL, Maibach HI (1985) Percutaneous absorption of nitroaromatic compounds: in vivo and in vitro studies in human and monkey skin. J Invest Dermatol 64:190–195
Bronaugh RL, Maibach HI (1999) Percutaneous absorption, 3rd edn. Marcel Dekker, Inc. CRC Press, New York
Buchwald P, Bodor N (1998) Octanol-water partitioning: searching for predictive models. Curr Med Chem 5:353–380
Buchwald P, Bodor N (2001) A simple, predictive structure-based skin permeability model. J Pharm Pharmacol 53:1087–1098
Chilcott RP, Barai N, Beezer AE, Brain SL, Brown MB, Bunge AL, Burgess SE, Cross S, Dalton CH, Dias M, Farinha A, Finnin BC, Gallagher SJ, Green DM, Gunt H, Gwyther RL, Heard CM, Jarvis CA, Kamiyama F, Kasting GB, Ley EE, Lim ST, McNaughton GS, Morris A, Nazemi MH, Pellett MA, Du Plessis J, Quan YS, Raghavan SL, Roberts M, Romonchuk W, Roper CS, Schenk D, Simonsen L, Simpson A, Traversa BD, Trottet L, Watkinson A, Wilkinson SC, Williams FM, Yamamoto A, Hadgraft J (2005) Inter- and intra-laboratory variation of in vitro diffusion cell measurements: an international multicenter study using quasi-standardised methods and materials. J Pharm Sci 94:632–638
Clowes HM, Dick IP, Thornley K, Scott RC (1993) In vitro percutaneous absorption measurements: Flow-through or static diffusion cells? In: Brain KR, James VJ, Walters KA, (eds) Prediction of percutaneous penetration, vol 3b. STS Publishing, Cardiff, pp 117–122
Dal Pozzo AD, Donzelli G, Liggeri E, Rodriguez L (1991) Percutaneous absorption of nicotinic acid derivatives in vitro. J Pharm Sci 80:54–57
Dearden JC, Cronin MTD. Patel H., Raevsky OA (2000) QSAR prediction of human skin permeability coefficients. J Pharm Pharmacol 52S:221
El Tayar N, Tsai RS, Testa B, Carrupt PA, Hansch C, Leo A (1991) Percutaneous penetration of drugs—a quantitative structure-permeability relationship study. J Pharm Sci 80:744–749
Flynn GL (1990) Physicochemical determinants of skin absorption. In: Gerrity TR, Henry CJ (eds) Principles of route-to-route extrapolation for risk assessment. Elsevier, New York, pp 93–127
Franz TJ (1975) Percutaneous absorption: on the relevance of in vitro data. J Invest Dermatol 64:190–195
Friend DR (1992) In vitro skin permeation techniques. J Cont Rel 18:235–248
Geinoz S, Guy R, Testa B, Carrupt P (2004) Quantitative structure-permeation relationships (QSPeRs) to predict skin permeation: a critical evaluation. Pharm Res 21:83–92
Ghafourian T, Fooladi S (2001) The effect of structural QSAR parameters on skin penetration. Int J Pharm 217:1–11
Ghafourian T, Safari A, Adibkia K, Parviz F, Nokhodchi A (2007) A drug release study from hydroxypropylmethylcellulose (HPMC) matrices using QSPR modelling. J Pharm Sci 96:3334–3351

Ghafourian T, Samaras EG, Brooks JD, Riviere JE (2010) Modelling the effect of mixture components on permeation through skin. Int J Pharm 398:28–32

Goodman M, Barry BW (1988) Action of penetration enhancers on human skin as assessed by the permeation of model drugs 5-fluorouacil and estradiol 1. Infinite dose technique. J Invest Dermatol 91:323–327

Grissom RE, Brownie C, Guthrie FE (1987) In vivo and in vitro dermal penetration of lipophilic and hydrophilic pesticides in mice. Bull Environ Contam Toxicol 38:917–924

Guha R, Jurs PC (2005) Interpreting computational neural network QSAR models: a measure of descriptor importance. J Chem Inf Mod 45:800–806

Hadgraft J, Ridout G (1987) Development of model membranes for percutaneous absorption measurements 1. Isopropyl myristate. Int J Pharm 39:149–156

Hewitt M, Binjumah WMS, Cronin MTD, Gullick DR, Sun Y, Davey N, Wilkinson SC, Moss GP (2014) QSAR and machine learning analysis of a PDMS dataset. In: Brain KR, Walters KA (eds) Perspectives in percutaneous penetration, vol 14. STS Publishing, Cardiff, p 87

Hotchkiss SAM, Chidgey MAJ, Rose S, Caldwell J (1990) Percutaneous absorption of benzyl acetate through rat skin in vitro. 1. Validation of an in vitro model against in vivo data. Food Chem Toxicol 1990(28):443–448

Hou SYE, Flynn GL (1989) Enhancement of hydrocortisone permeation of human and hairless mouse skin by 1-dodecylazacycloheptan-2-one. J Invest Dermatol 93:774–779

Johnson ME, Blankstein D, Langer R (1995) Permeation of steroids through human skin. J Pharm Sci 84:1144–1146

Kasting GB, Smith RL, Cooper ER (1987) Effect of lipid solubility and molecular size on percutaneous absorption. Skin Pharmacol 1:138–153

Kasting GB, Filloon TG, Francis WR, Meredith MP (1994) Improving the sensitivity of in vitro skin penetration experiments. Pharm Res 1994(11):1747–1754

Knutson K, Harrison DJ, Pershing LK, Goates CY (1993) Transdermal absorption of steroids. J Cont Rel 24:95–108

Lam LT, Sun Y, Davey N, Adams RG, Prapopoulou M, Brown MB, Moss GP (2010) The application of feature selection to the development of Gaussian process models for percutaneous absorption. J Pharm Pharmacol 62:738–749

Lin RY, Hsu CW, Chen WY (1996) A method to predict the transdermal permeability of amino acids and dipeptides through porcine skin. J Cont Rel 38:229–234

Liu PC, Kuriharabergstrom T, Good WR (1991) Co-transport of estradiol and ethanol through human skin in vitro—understanding the permeant enhancer flux relationship. Pharm Res 8:938–944

Lovering EG, Black DB (1974) Diffusion layer effects on permeation of phenylbutazone through polydimethylsiloxane. J Pharm Sci 63:1399–1402

Magnusson BM, Anissimov YG, Cross SE, Roberts MS (2004) Molecular size as the main determinant of solute maximum flux across the skin. J Invest Dermatol 122:993–999

Michaels AS, Chanderasekaran SK, Shaw JE (1975) Drug permeation through human skin: theory and in vitro experimental measurement. AIChE J 21:985–996

Mistry T, Ajum K, Owen J, Stair J, Wilkinson SC, Staff K, Moss GP (2013) The percutaneous absorption of aluminium from antiperspirant products. In: Brain KR, Chilcott R (eds) Advances in the dermatological sciences. Royal Society of Chemistry, Cambridge, pp 256–264

Mitragotri S, Edwards DA, Blankstein D, Langer R (1995) Mechanistic study of ultrasonically-enhanced transdermal drug delivery. J Pharm Sci 84:697–706

Mitragotri S, Anissimov YG, Bunge AL, Frasch HF, Guy RH, Hadgraft J, Kasting GB, Lane ME, Roberts MS (2011) Mathematical models of skin permeability: an overview. Int J Pharm 418:115–129

Moss GP (2014) All models are wrong, but some are useful—linear and non-linear models for percutaneous absorption. In: Brain KR, Walters KA (eds) Perspectives in percutaneous penetration, vol 14. STS Publishing, Cardiff, p 6

Moss GP, Cronin MTD (2002) Quantitative structure-permeability relationships for percutaneous absorption: re-analysis of steroid data. Int J Pharm 238:105–109
Moss GP, Gullick DR, Pugh WJ (2005) Do QSPR models accurately reflect percutaneous absorption? J Pharm Pharmacol 57S:18
Moss GP, Gullick DR, Cox PA, Alexander C, Ingram MJ, Smart JD, Pugh WJ (2006) Design, synthesis and characterisation of captopril prodrugs for enhanced percutaneous absorption. J Pharm Pharmacol 58:167–177
Moss GP, Shah AJ, Adams RG, Davey N, Wilkinson SC, Pugh WJ, Sun Y (2012) The application of discriminant analysis and Machine Learning methods as tools to identify and classify compounds with potential as transdermal enhancers. Euro J Pharm Sci 45:116–127
Moss GP, Wilkinson SC, Kyriacou T, Gullick DR (2013) NARMAX models in the prediction of penetration across skin and polydimethylsiloxane membranes. In: Brain KR, Chilcott R (eds) Advances in the dermatological sciences. Royal Society of Chemistry, Cambridge, pp 384–388
Neumann D, Kohlbacher O, Merkwirth C, Lengauer T (2006) A fully computational model for predicting percutaneous drug absorption. J Chem Inf Mod 46:424–429
Nugent FE, Wood JA (1980) Methods for the study of percutaneous absorption. Can J Pharm Sci 15:12–17
Patel H, ten Berge W, Cronin MTD (2002) Quantitative structure-activity relationships (QSARs) for prediction of skin permeation of exogenous chemicals. Chemosphere 48:603–613
Potts RO, Guy RH (1992) Predicting skin permeability. Pharm Res 9:663–669
Ranadive SA, Chen AX, Serajuddin ATM (1992) Relative lipophilicities and structural pharmacological considerations of various angiotensin-converting enzyme ACE inhibitors. Pharm Res 9:1480–1486
Reifenrath WG, Lee B, Wilson DR, Spencer TS (1994) A comparison of in vitro penetration cells. J Pharm Sci 83:1229–1233
Sartorelli P, Aprea C, Cenni A, Novelli MT, Orsi D, Palmi S, Matteeucci G (1998) Prediction of percutaneous absorption from physicochemical data: a model based on data of in vitro experiments. Ann Occ Hyg 42:267–276
Scheuplein RJ, Blank IH, Brauner GI, MacFarlane DJ (1969) Percutaneous absorption of steroids. J Invest Dermatol 52:63–70
Scott RC, Batten PL, Clowes HM, Jones BK, Ramsey JD (1992) Further validation of an in vitro method to reduce the need for in vivo studies for measuring the absorption of chemicals through the skin. Fundam Appl Toxicol 19:484–492
Southwell D, Barry BW, Woodford R (1984) Variations in permeability of human skin within and between specimens. Int J Pharm 18:299–309
Stehle RG, Higuchi WI (1972) In vitro model for transport of solutes in three-phase systems II: experimental considerations. J Pharm Sci 61:1931–1935
Sun Y, Moss GP, Davey N, Adams R, Brown MB (2011) The application of stochastic Machine Learning methods in the prediction of skin penetration. App Soft Comput 11:2367–2375
van de Sandt JJM, Maas WJM, van Burgsteden JA, Sartoreilli P, Montomoli L, Larese F, Payan J-P, Limasset JC, Carmichael P, Kenyon S, Robinson E, Dick I, Nielsen JB, Schaller K-H, Korinth G, Geh S, Cage S, Wilkinson S, Williams FM (2002) International evaluation of the robustness of in vitro percutaneous absorption studies. In: Brain KR, Walters KA (eds) Perspectives in percutaneous penetration, vol 8a. STS Publishing, Cardiff, p 126
Wagner H, Kostka K-H, Lehr C-M, Schaefer UF (2000) Drug distribution in human skin using two different in vitro test systems: comparison with in vivo data. Pharm Res 2000(17): 1475–1481
Walters KA, Brain KR (2000) Skin permeation predictions and databases—what are their limitations? In: Walters KA, Brain KR (eds) Perspectives in percutaneous penetration, Vol 7a. 7th international conference, La Grande Motte, France, p 15
Walters KA, Brain KR, Howes D, James VJ, Kraus AL, Teetsel NM, Toulon M, Watkinson AC (1997) Percutaneous penetration of octyl salicylate from representative sunscreen formulations through human skin in vitro. Food Chem Tox 35:1219–1225

Williams AC (2003) Transdermal and topical drug delivery. The Pharmaceutical Press, London
Williams AC, Cornwell PA, Barry BW (1992) On the non-Gaussian distribution of human skin permeabilities. Int J Pharm 86:69–77
Wilschut A, ten Berge WF, Robinson PJ, McKone TE (1995) Estimating skin permeation: the validation of five mathematical models. Chemosphere 30:1275–1296
Yum S, Lee E, Taskovich L, Theeuwes FI (1994) In: Hsieh DS (ed) Drug permeation enhancement: theory and application. Marcel Dekker, New York, pp 143–170

Chapter 10
Conclusions and Recommendations for Model Development and Use

Overview of the Previous Chapters

This book has reviewed the main achievements in the development of predictive algorithms for percutaneous absorption. As mentioned in Chap. 9, it is not an exhaustive review and clearly reflects, in some cases, the research interests of the authors. The aims of this chapter are to draw together and summarise the previous discussions, and to conclude with a series of recommendations which outline how models should be developed and reported.

The starting point, as with any aspect of percutaneous absorption, is the physiology of the skin (Chap. 1). While the main barrier to percutaneous absorption is known to lie predominately in the outermost layer, the *stratum corneum*, the permeation process is complicated by the comparatively hydrophilic nature of the underlying viable epidermis and dermis, which mitigates the absorption of appreciable amounts of highly lipophilic compounds beyond the *stratum corneum*.

Chapter 2 described how in vitro measurements of percutaneous absorption are conducted. It focused on Franz-type cell apparatus as this is the most commonly used type of diffusion cell, and it is from cells of this type that the majority of data used to develop models have been obtained. Thus, the benefits and issues with such diffusion cells were discussed, including the experimental temperature, donor and receptor solvent choices, whether the donor compartment is occluded or unoccluded and how sink conditions are maintained in order to provide as realistic a model of permeation as possible. Thus, in Chap. 9, these methods are discussed in the context of their impact on the quality of any models derived from such experimental studies.

Chapter 3 focuses on the underlying theories of skin absorption and on theoretical aspects of permeant clearance (i.e. pharmacokinetics), describing the most practicable and widely used version of Fick's law to quantify usable parameters—flux (as J, J_{max} or $J_{steadystate}$) and the permeability coefficient (k_p). It then examined in detail the key initial works in this field (Flynn 1990; El Tayar et al. 1991;

G.P. Moss et al., *Predictive Methods in Percutaneous Absorption*,
DOI 10.1007/978-3-662-47371-9_10

Potts and Guy 1992) which was followed in Chap. 4 by a detailed review of subsequent research by a number of researchers, including the modifications of Cleek and Bunge (Cleek and Bunge, 1993; Bunge and Cleek 1995; Bunge et al. 1995) to Potts and Guy's model, the second Potts and Guy (1995) model examining only non-electrolytes from Flynn's data set and Wilschut's consideration of a range of models (Wilschut et al. 1995), some of which included nonlinear descriptors. Chapter 5 focused on models developed for artificial membranes, most notably polydimethylsiloxane, which have been used as a substitute for mammalian skin in in vitro experiments.

Chapter 6 reviewed some of the more novel research in this field that sits somewhat outside the mainstream approaches. In particular, the work of Magnusson et al. (2004a, b) and subsequent research by this group are highlighted due to its novel approach and burgeoning significance in the field. Chapter 7 discusses not only the benefits but also the significant current limitations of machine learning methods applied to this problem domain, while Chap. 8 describes the application of modelling methods to finite-dose systems and also considers formulation aspects, focusing on the work of Ghafourian and Riviere. Chapter 9 presents a critical review of the models developed, focusing on experimental issues and how they impact on the development, validation and accuracy of models, as well as how models have been misinterpreted or used beyond their appropriate "chemical space".

This chapter therefore considers some of the key recommendations in the wider field of QSAR (and similar) model development and how those recommendations can be realistically applied to the development of quantitative models of skin absorption so that any resultant models are relevant, as transparent as possible and of practical benefit to physical scientists.

"Pitfalls" of Model Development and Use

The title of this section is borrowed from a landmark paper by Cronin and Schultz (2003) which will form the initial aspect of this discussion. Their study focused broadly on the development of QSAR models in many fields, not just skin permeation, and this section will discuss how their recommendations may, or may not, apply to skin permeation. They commented that the field of QSAR (and similar) model development has grown over the last twenty years or so without either a defined framework, series of rules for model development and validation or guidelines for methodology. As an overview, they stated that the aim of any such quantitative model is to develop models based on a training set of compounds. This then allows for the prediction of biological activity of related chemicals—related, in a general sense, to the boundaries of the data as defined by, for example, Potts and Guy (1992) where their model featured data where $-3 < \log P < 6$ and $150 < MW < 750$, and which defined the limits of the model based on the input, or training, data set. Further, the use of multiple linear regression, or related, methodology ensures that the models are transparent and provide mechanistic insight into the biological process being modelled. Cronin and Schultz commented that the use of such methods has potential limitations that are often not appreciated or which are not reported by those developing QSARs. They therefore listed a series of "essential" and "desirable" qualities which they concluded QSARs should exhibit, and discussed these issues in the context of potential pitfalls in model design and use, where they used many examples of their own work to suggest better practice in developing models.

Their framework was based around three main components associated with models and described throughout this book in substantial detail. They are the biological endpoint data (in the case of skin permeability, this is normally the permeability coefficient or, in some cases, flux), structural or physicochemical properties that describe a molecule and the statistical technique used to relate the structure to the biological property. A summary of their conclusions, modified to contextualise their comments for percutaneous absorption, is shown in Table 10.1.

Table 10.1 Essential and desirable features for a quantitative model for percutaneous absorption (modified from Cronin and Schultz 2003)

Essential properties	Desirable properties
A well-defined and readily measurable endpoint (i.e. permeability, k_p, or flux (J_{max} or J_{ss})	Do not extrapolate or use the model beyond its original domain, both in terms of the range of physicochemical descriptors used or in terms of the class(es) of chemicals from which the model was developed
A data set that is chemically and biologically diverse and that is equally representative of the "chemical space" associated with the model; the data set should be split into suitably representative training and test sets, which are separate and consistent	The model should be correctly used and interpreted within the context of the model's findings and of prior knowledge in the field
Physiochemical descriptors that are, in number and type, consistent with the skin permeability endpoint (i.e. molecular features that have been shown experimentally or empirically to be significant in the permeation process)	Models and methodologies should be transparent and not based on "black box" approaches
The method of analysis should be appropriate and produce a model that is statistically valid; statistical performance should be described for each mode	Appreciation of the precision of the model when interpreting its output (in the case of models featuring algorithms); mean and standard deviation values for each significant descriptor should be provided
The model and method should have a strong mechanistic basis	The use of a single model to describe a process, rather than a series of models to describe individual parts of the process
Development of models by multidisciplinary groups of experts	

Quality of the Source, or Input, Data

The biological data for skin absorption, upon which current models are based, are usually either permeability or flux. This should not, however, rule out future developments where different endpoints may be used to predict optimum absorption in, for example, superficial tissues to improve the topical delivery of a range of drugs, including for example antifungals and anti-irritants. In contrast to the general picture presented by Cronin and Schultz, the field of percutaneous absorption uses well-defined measures of biological data (permeability data). The underlying issue that is more specific to studies of skin absorption is in the quality of the data used; in particular, the consistency of experimental protocols, which include everything from selection and quality of skin to the choice of formulation and receptor solvents, is central to the development of models of skin absorption. While understanding that a model should only be applied within its limits—including its experimental limits—a key recommendation for the future development of models of skin permeability is that they adopt a standardised protocol. Ideally, this would relate to a single study, which is unlikely, and the next best option—a consideration of the approach proposed by Chilcott et al. (2005) has been shown to provide little improvement in data quality or reproducibility. Nevertheless, as issues with Flynn's (1990) data set have shown (i.e. Johnson et al. 1995; Moss and Cronin 2002) a reduction in the number of sources of data from which a model is constructed would appear to be central to producing a valid and reliable model. Thus, permeability data from which models are constructed should ideally come from a single protocol—and from the same laboratory, or workers, if possible—to reduce potential sources of error. The protocol should be well established and validated with clear and well-defined endpoints. As Cronin and Schultz pointed out, there are numerous examples of good practice in other fields that those developing models of percutaneous absorption could learn from in this regard; in these cases, the underlying issue is the development of an open-access database resource to which data points are added only if they meet the criteria described above. However, given that a large amount of the data already used for the development of models of skin permeability use published data from a variety of sources, this will—at worst—result in data that are not comparable and—at best—that exhibit substantial variation. While such data could be used for modelling, it should be used with caution, and an understanding of the limitations placed upon it by its source.

Such restrictions will be reflected in the poor statistical fit of any models thus derived. A number of suggestions for using data of perceived poor quality were discussed by Cronin and Schultz. The modeller should consider the nature of the data set, particularly its source(s) and quality, when considering the statistical quality of the resultant model. Models should, where possible, be developed empirically and be pragmatic; that is, models can be based on a small number of parameters of known significance from previous studies—perhaps echoing the use of log P and molecular weight by Flynn, and hydrogen bonding by Roberts and colleagues (Roberts et al. 1995, 1996; Pugh et al. 1996, 2000)—rather than analysing a large number of descriptors which are known to be largely irrelevant. They also recommend robust validation of the models using distinct training and test data sets.

Outliers

Cronin and Schultz also discussed outliers—compounds poorly predicted by a model—in the development of QSARs. Generally, outliers are identified by their lack of statistical fit, and it is usually inferred that this relates to a different mechanism of action than the rest of the data analysed that are not characterised as outliers. In the context of predictive models for skin permeation, the clear example of the revised Scheuplein data, which comprised almost 15 % of Flynn's original data set, is significant (Scheuplein et al. 1969). This has been discussed previously in Chap. 4 but highlights the issue not only with identifying and selecting outliers, but also with the chosen method of analysis, which may result in misleading methods being used and which highlights the recommendations of Moss et al. (2009) to undertake rudimentary analyses of the data sets in order to characterise their fundamental nature (i.e. whether the data follows linear or nonlinear trends) so that the correct methods of analysis can be chosen.

Methods to highlight outliers include their identification based on their high standard residuals from regression-based techniques; following this, they are often removed individually either by subjective comment—perhaps informed by empirical insights of how, for example, a particular chemical might permeate the skin—or by whether they sit above or below an arbitrary cut-off point (i.e. a particular residual value returned from statistical analysis). When carried out correctly, the removal of outliers—and the identification of which chemicals were removed for this reason—will improve the quality and relevance of a model. In some cases, it may be relevant to analyse a model both before and after the removal of outliers, as those compounds which are genuine outliers will, if removed, result in a minimal change to the model. However, the situation with Johnson et al's (1995) reanalysis of the Flynn data set, which highlights issues of data quality, should be borne in mind when considering such manipulations of the model.

Biological Data

Cronin and Schultz make the obvious, but often overlooked, comment that biological data are inherently variable and subject to error and that standard protocols may often be difficult to develop. In the case of skin absorption, while it is reasonable to comment that methods for the measurement of in vitro percutaneous absorption are reasonably standardised, significant differences do exist. As such experiments are the main source of data from which models are constructed, it is difficult to remove such variation from models of skin permeation. Therefore, models of skin permeation should be presented as mean values from a series of replicates, with the standard deviation or standard error also quoted—a good example of this practice is shown in the review by Geinoz et al. (2004). Thus, an extension of this consideration is that, once it is accepted that biological measurements are associated with error, it should be considered that certain protocols might result in more error than others. This may be considered when collating literature data into a single data set.

Descriptor Selection and Interpretation, and Data Set Design

In selecting which descriptors to use for an analysis, care should be taken to avoid repetition, colinearity and therefore relevance, which has been observed previously with topological indices such as molecular connectivity (Basak et al. 2000; Patel and Cronin 2001). To avoid such issues, any regression analysis that is based on multivariate methods should not be based on colinear descriptors, as this will result in an artificially high regression coefficient (Romanelli et al. 2000). This is achieved by the analysis of the correlation matrix output from any regression analysis, but decisions based on what is or is not an acceptable level of covariance are somewhat subjective but should be as low as possible, but must be significantly lower than the statistical fit of the model itself, and the r^2 value (adjusted for degrees of freedom) should be reported (Cronin and Schultz 2003).

While poor data set design can result in issues of colinearity and introduce bias into the data it is, in the case of percutaneous absorption, extremely difficult to find sufficient data in the literature to compile a data set that is completely without some form of bias. In particular, most data relate to chemicals that have low to intermediate lipophilicities (i.e. $1.0 < \log P < 3.0$, or MW < 500) and, as such, the range and relevance of any resultant models may be of limited value—this is discussed in detail in Chap. 9 (Moss et al. 2006).

More broadly, poor data set design may result in the inclusion of colinear descriptors which are of little relevance to the underlying mechanism of the process being modelled. This is more common with related chemicals (i.e. homologous series) and may be addressed by using as diverse a data set as available. This is, of course, somewhat idealistic, but it does highlight the relevance of models of skin permeability and their limitations. Statistical methods, including the use of principal component analysis (Moss et al. 2009; Sun et al. 2011) or correlation matrices and an examination of the intercorrelations between variables, should be assessed in any method to which such statistical measures are relevant. Thus, issues of colinearity may be reduced by the selection of relevant fundamental physicochemical descriptors which will allow clear and unambiguous mechanistic inferences to be drawn.

Statistical Analysis of Data

Although Cronin and Schultz (2003) focused on QSAR-based modelling approaches, their comments on statistical analysis are broadly relevant and impact on considerations of other methods. Over-fitting of data may be an issue which is related to the method of analysis, and potentially also of the variance associated with data of a biological origin. This is also an issue for nonlinear methods and was discussed in Chap. 7, where artificial neural networks in particular have been shown to over-fit data.

Nevertheless, most biological processes are inherently nonlinear, and, in the context of fundamental physicochemical parameters, it is clear that this is the case with skin absorption where both highly hydrophilic and highly hydrophobic chemicals are poor skin permeants. Thus, Cronin and Schultz commented that global modelling is unlikely to successful without some consideration of nonlinearity. This is perhaps reflected in the improved models obtained—compared to linear methods—when various researchers have employed nonlinear methods. These methods have been discussed in Chaps. 4–7.

However, one issue with some nonlinear methods—discussed in Chap. 7—is their inherent lack of transparency. Thus, while machine learning and related methods currently appear to offer improved predictions of percutaneous absorption, they lack transparency and have limited portability. It may therefore be the case that despite advances described in Chap. 7, the modeller should use the most transparent, portable and readily interpretable method available as it may offer greater utility, particularly in terms of mechanistic insight, than "better" models.

Another issue with predictive models is that they are often used outside their "chemical space"; that is, predictions are made outside the area of knowledge covered by the training set. Quite often, the training set is not defined in the model and to do so would allow transparent and relevant predictions to be made from models. However, this is complicated by the issue of the "chemical space" of models and how even this poorly fits the predictions of the model—this is discussed in detail in Chap. 9. Cronin and Schultz commented that the reporting of the model and its statistical nature, range of data and other specific features (i.e. relevant commentary on colinearity) should be presented to avoid ambiguity and misuse of models.

In discussing "new" methods, such as those based on machine learning methods, Cronin and Schultz's comments are highly significant. They commented on the development of QSARs—and, by extension, other methods—in developing relationships between the physiochemical properties of a chemical and a particular biological response in the context of multidisciplinary research groups whose expertise extends across all parts of the study and its methodology. In essence, this implies that, rather than simply collecting data from the literature, researchers in percutaneous absorption need to refine their data selection based on the issues relating to the quality of the data available. The relevance of Cronin and Schultz's wok has been highlighted as it has informed the OECD principles for the validation of (Q)SAR Models[1] (OECD 2008).

Data—and Data Set—Quality

This is particularly important in considering the size of the data set and the significance of analyses related to subsets abstracted from larger sources of data, which themselves may be comprised of subsets taken from different sources, where both subsets may not necessarily be the same. While Cronin and Schultz discuss the quality of data, they do so broadly for a range of fields and scientific disciplines. In the context of the comments raised by Moss et al. (2002), discussed above, the quality of the model is seldom discussed in developing quantitative models for skins absorption, other than a brief description of the sources of the data, and it is often inferred that the data are "acceptable" or that the reader will have explored the cited sources further.

The number of data points used to develop the model is also significant. There may be benefits in extrapolating a model beyond its range, but this may provide significantly limited information due to the definition of "range", which can include both the range of physicochemical descriptors and the diversity of the chemicals included in the data set. This latter point may impact on the development of a data set free from bias (i.e. skew) and was discussed by Sun et al. (2011), who demonstrated how model quality varied due to increases in covariance as models extended outside highly populated parts of their "chemical space"—in this case, it is important to note that the performance of the model was variable *within its defined "chemical space"*, a point seldom discussed and which is significant to the overall performance of the model, which might be affected by issues such as covariance. This was also highlighted by Moss et al. (2006), who demonstrated the poor predictions produced by a range of linear models at the limit of their "chemical space". Thus, similar performance issues might be expected when using the preponderance of linear algorithms at the limits of their ranges.

Chapters 4 and 5 have discussed the development of quantitative models from a range of small data sets, with some containing as few as four chemicals. Clearly, the quality and relevance of such models is questionable, being limited by the volume of amount of data available (Moss et al. 2011). On the other hand, such studies and their limitations do emphasise the underlying issue of data availability and might, in the case of small data sets taken from single sources, be argued to actually improve model quality due to an improved consistency of data. The high statistical quality of some of these models, particularly their r^2 values, may suggest over-fitting. It is also

[1]Available at http://www.oecd.org/document/4/0,3746,en_2649_34379_42926724_1_1_1_1,00.html.

Table 10.2 Types of error associated with the development and use of QSAR/QSPR models (modified from Dearden et al. 2007)

Number	Type of error	Associated OECD principle(s) (OECD 2008)
1	Failure to account for data heterogeneity	1
2	Use of inappropriate endpoint data	1
3	Use of colinear descriptors	2, 4, 5
4	Use of incomprehensible descriptors	2, 5
5	Error in descriptor values (from experimental or computed sources)	2
6	Poor transferability of the model	2
7	Inadequate or ill-defined domain of applicability	3
8	Unacknowledged omission of data points, either through statistical analysis (i.e. removal of outliers) or empirically	3
9	Use of inadequate data	3
10	Replication of the same data points (multiple values of the same chemical) in a data set	3
11	A range of endpoint (permeability or flux) data that are too narrow	3
12	Over-fitting of data	4
13	Use of excessive number of descriptors without justification (i.e. feature selection methods)	4
14	Lack of an inadequate description of the statistical nature of the model	4
15	Incorrect calculation	4
16	Lack of descriptor auto-scaling	4
17	Misuse or misinterpretation of statistics	4
18	No consideration of the distribution of residuals	4
19	Selection of test and training sets that are not separate but equally diverse and representative of the process being modelled	4
20	Lack of model validation	4
21	Lack of mechanistic interpretation	5

interesting to reflect that these models based on subsets are often substantially different from both each other and the larger parent data set from which they were abstracted.

Thus, the size of the data set used, in both absolute and comparative senses, influences model quality significantly and has produced counterintuitive outputs in many cases, and results did not agree with laboratory studies (Moss et al. 2006, 2011). However, issues of the data set design, including the use of representative and unbiased data, are significant, but other issues, such as the use of computed or experimentally derived parameters—or a mixture of both—have not been fully addressed in such studies, nor has the issue that "computed" values, particularly melting point and the octanol–water partition coefficient (log *P*), may often be computed by a range of different methods if taken from different sources and collated into a single data set. It should also be noted that recent work on improving the analysis of small data sets may provide significant advances in this field (Ashraft et al. 2015).

Thus, in using the available skin permeability data, the resultant models are significantly limited, even when they produce highly specific and insightful outputs, such as the set of "simple rules" for percutaneous absorption, defined by Magnusson et al. (2004a, b). They concluded—using data from the literature—that good skin penetrants had MW $\leq$ 152, aqueous solubility (log *S*) > −2.3, number of atoms available for hydrogen bonding (HB) $\leq$ 5, octanol–water partition coefficient (log *K*) < 2.6 and melting point (MPt, °K) $\leq$ 432. In contrast, poor penetrants have MW > 213, log *S* < −1.6, HB $\geq$ 4, log *K* > 1.2 and MPt $\geq$ 223. Discriminant analysis demonstrated that 70 % of chemicals, based on the above descriptors, could be successfully assigned. Thus, while this simple and elegant study enables rapid initial screening of potential permeants, for either drug delivery or environmental risk assessment, it does so in the context of the underlying issues associated with the source data described herein. Thus, there is only so far such a model can go in terms of accurately defining such phenomena. Dearden et al. (2007) defined 21 sources of error associated variously with the thousands of quantitative structure (and permeability) relationship models published. These are summarised in Table 10.2 and provide an excellent guide to the sensible and useful development of quantitative models. In doing so, however, Dearden et al. expose the weakness of the foundations of quantitative models of skin absorption but also suggest clearly—through the development of consistent, transparent models based on validated data—how improved models may be developed in the future.

Conclusions

Thus, the significance of how the input affects the output, or—to use a crude term from the field of computing—"garbage in, garbage out", cannot be overstated. This implies that the real advantages of advanced methods, such as artificial neural networks and machine learning methods, may not be fully appreciated with the current availability of data. The field of modelling percutaneous absorption is still in its early stages, and there is still a paucity of consistent, high-quality data from which models can be developed. The endpoints of models may also change, as some models may be specifically focused on particular chemicals or a particular biological process, and this will expand the application of these models to considerations of dermal absorption to other processes, such as topical deposition and localised therapies in the skin at the site of application. The work undertaken in other fields, particularly ecotoxicology, provides a successful template for the potential harmonisation of models and methodologies on which researchers can, in the future, build models with greater predictive ability and mechanistic relevance.

References

Ashraft P, Sun Y, Davey N, Adams R, Moss G, Brown MB, Prapopoulou M (2015) The importance of hyperparameter selection within small datasets. In: International Joint Conference on Neural Networks, Killarney, Ireland, p 139. Available at http://www.ijcnn.org/assets/docs/ijcnn2015-program-v3.pdf. Accessed 7 June 2015

Basak SC, Balaban AT, Grunwald GD, Gute BD (2000) Topological indices: their nature and mutual relatedness. J Chem Inf Comput Sci 40:891–898

Bunge AL, Cleek RL (1995a) A new method for estimating dermal absorption from chemical exposure 2 effect of molecular weight and octanol-water partitioning. Pharm Res 12, 88–95

Bunge AL, Cleek RL, Vecchia BE (1995b) A new method for estimating dermal absorption from chemical exposure 3 compared with steady-state methods for prediction and data analysis. Pharm. Res 12:972–982

Chilcott RP, Barai N, Beezer AE, Brain SL, Brown MB, Bunge AL, Burgess SE, Cross S, Dalton CH, Dias M, Farinha A, Finnin BC, Gallagher SJ, Green DM, Gunt H, Gwyther RL, Heard CM, Jarvis CA, Kamiyama F, Kasting GB, Ley EE, Lim ST, McNaughton GS, Morris A, Nazemi MH, Pellett MA, Du Plessis J, Quan YS, Raghavan SL, Roberts M, Romonchuk W, Roper CS, Schenk D, Simonsen L, Simpson A, Traversa BD, Trottet L, Watkinson A, Wilkinson SC, Williams FM, Yamamoto A, Hadgraft J (2005) Inter- and intra-laboratory variation of in vitro diffusion cell measurements: an international multicenter study using quasi-standardised methods and materials. J Pharm Sci 94:632–638

Cleek RL, Bunge AL (1993) A new method for estimating dermal absorption from chemical exposure 1 general approach. Pharm Res 10:497–506

Cronin MTD, Schultz, WT (2003) Pitfalls in QSAR. J Theoret Chem (Theochem). 622:39–51

Dearden JC, Cronin MTD, Kaiser KLE (2007) How not to develop a quantitative structure-activity or structure-property relationship (QSAR/QSPR). SAR QSAR Environ Res 20:241–266

El Tayar N, Tsai RS, Testa B, Carrupt PA, Hansch C, Leo A (1991) Percutaneous penetration of drugs—a quantitative structure-permeability releationship study. J Pharm Sci 80:744–749

Flynn GL (1990) Physicochemical determinants of skin absorption. In: Gerrity TR, Henry CJ (eds) Principles of route-to-route extrapolation for risk assessment. Elsevier, New York, pp 93–127

Geinoz S, Guy R, Testa B, Carrupt P (2004) Quantitative structure-permeation relationships (QSPeRs) to predict skin permeation: a critical evaluation. Pharm Res 21:83–92

Johnson ME, Blankstein D, Langer R (1995) Permeation of steroids through human skin. J Pharm Sci 84:1144–1146

Magnusson BM, Anissimov YG, Cross SE, Roberts MS (2004a) Molecular size as the main determinant of solute maximum flux across the skin. J. Invest. Dermatol. 122:993–999

Magnusson BM, Pugh WJ, Roberts MS (2004b) Simple rules defining the potential of compounds for transdermal delivery or toxicity. Pharm Res 21:1047–1054

Moss GP, Cronin MTD (2002) Quantitative structure-permeability relationships for percutaneous absorption: re-analysis of steroid data. Int J Pharm 238:105–109

Moss GP, Dearden JC, Patel H, Cronin MTD (2002) Quantitative structure-permeability relationships (QSPRs) for percutaneous absorption. Tox In Vitro 16:299–317

Moss GP, Gullick DR, Cox PA, Alexander C, Ingram MJ, Smart JD, Pugh WJ (2006) Design, synthesis and characterisation of captopril prodrugs for enhanced percutaneous absorption. J Pharm Pharmacol 58:167–177

Moss GP, Sun Y, Prapopoulou M, Davey N, Adams R, Pugh WJ, Brown MB (2009) The application of Gaussian processes in the prediction of percutaneous absorption. J Pharm Pharmacol 61:1147–1153

Moss GP, Sun Y, Wilkinson SC, Davey N, Adams R, Martin GP, Prapopoulou M, Brown MB (2011) The application and limitations of mathematical models across mammalian skin and poldimethylsiloxane membranes. J Pharm Pharmacol 63:1411–1427

OECD (2008) OECD Principles for the Validation of (Q)SARs. http://www.oecd.org/dataoecd/33/37/37849783.pdf (Last accessed 18 Dec 2014)

Patel H, Cronin MTD (2001) A novel index for the description of molecular linearity. J Chem Inf Comput 41:1228–1236

Potts RO, Guy RH (1995) A predictive algorithm for skin permeability: the effects of molecular size and hydrogen bond activity. Pharm Res 12:1628–1633

Potts RO, Guy RH (1992) Predicting skin permeability. Pharm Res 9:663–669

Pugh WJ, Degim IT, Hadgraft J (2000) Epidermal permeability—penetrant structure relationships 4 QSAR of permeant diffusion across human stratum corneum in terms of molecular weight, H-bonding and electronic charge. Int J Pharm 197:203–211

Pugh WJ, Roberts MS, Hadgraft J (1996) Epidermal permeability—penetrant structure relationships 3 The effect of hydrogen bonding interactions and molecular size on diffusion across the *stratum corneum*. Int J Pharm 138:149–165

Roberts MS, Pugh WJ, Hadgraft J, Watkinson A (1995) Epidermal permeability—penetrant structure relationships 1 an analysis of methods of predicting penetration of monofunctional solutes from aqueous solutions. Int J Pharm 126:219–233

Roberts MS, Pugh WJ, Hadgraft J (1996) Epidermal permeability—penetrant structure relationships 2 the effect of H-bonding groups in penetrants on their diffusion through the *stratum corneum*. Int J Pharm 132:23–32

Romanelli GP, Cafferata LFR, Castro EA (2000) An improved QSAR study of toxicity of saturated alcohols. J Mol Struct (Theochem). 504:261–265

Scheuplein RJ, Blank IH, Brauner GI, MacFarlane DJ (1969) Percutaneous absorption of steroids. J Invest Dermatol 52:63–70

Sun Y, Moss GP, Davey N, Adams R, Brown MB (2011) The application of stochastic machine learning methods in the prediction of skin penetration. App Soft Comput 11:2367–2375

Wilschut A, ten Berge WF, Robinson PJ, McKone TE (1995) Estimating skin permeation: the validation of five mathematical models. Chemosphere 30:1275–1296

Index

Note: Page numbers followed by "f" and "t" indicate figures and tables respectively

G.P. Moss et al., *Predictive Methods in Percutaneous Absorption*,
DOI 10.1007/978-3-662-47371-9

MIX
Papier aus verantwortungsvollen Quellen
Paper from responsible sources
FSC® C105338

If you have any concerns about our products,
you can contact us on
ProductSafety@springernature.com

In case Publisher is established outside the EU,
the EU authorized representative is:
Springer Nature Customer Service Center GmbH
Europaplatz 3, 69115 Heidelberg, Germany

Printed by Libri Plureos GmbH
in Hamburg, Germany